The
Anthropology
of War

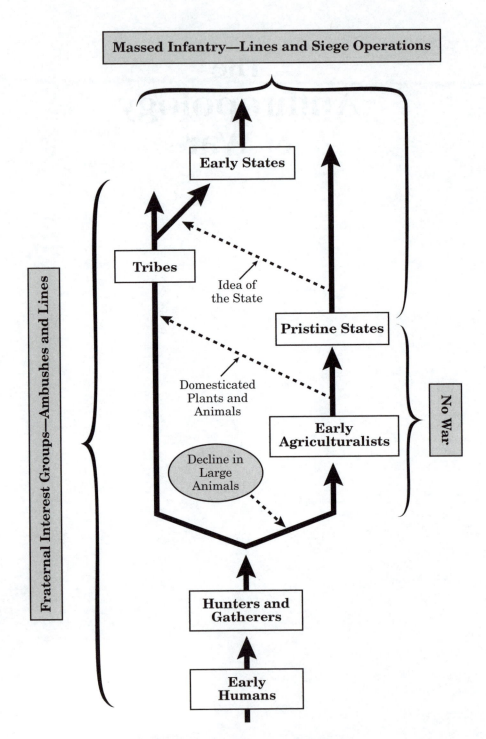

Massed Infantry—Lines and Siege Operations

Early States

Tribes

Idea of
the State

Pristine States

Domesticated
Plants and
Animals

**Early
Agriculturalists**

No War

Decline in
Large
Animals

Fraternal Interest Groups—Ambushes and Lines

**Hunters and
Gatherers**

**Early
Humans**

The Two Paths Warfare Has Taken

The Anthropology of War

Keith F. Otterbein
University at Buffalo

WAVELAND
PRESS, INC.
Long Grove, Illinois

For information about this book, contact:
Waveland Press, Inc.
4180 IL Route 83, Suite 101
Long Grove, IL 60047-9580
(847) 634-0081
info@waveland.com
www.waveland.com

10-digit ISBN 1-57766-607-0
13-digit ISBN 978-1-57766-607-3

Printed in the United States of America

7 6 5 4 3 2 1

To my son,
Gere Frederick Otterbein
(1971–2007)

Contents

Preface

This short book is an attempt to synthesize the findings of anthropologists of all specialties as well as those of some historians and political scientists. What you read here will differ in many respects from what you will find in my earlier and recent publications on warfare.

I read voraciously anything that I think might relate to the topic and include what I believe to be the most relevant information to fit within this book's space limitations. I used many means to locate materials: publishers' catalogues; book reviews; suggestions from friends, students, and relatives whom I have called or who have e-mailed me articles and references; walks through library stacks; my journal subscriptions; e-mail requests to scholars; online sources such as Amazon or Google; magazines; newspapers; radio (especially NPR); and TV (especially PBS and C-Span 2).

I also tried to respond to all requests and challenges to my work via letter, e-mail, or telephone calls.

Permission to reprint the frontispiece and figures 2.1 and 6.2 have been kindly granted by Texas A & M University Press. These figures are included in *How War Began* (2004).

I would like to single out the following individuals who have assisted me recently: Carol Berman, Richard Chacon, Robert Dentan, Brian Ferguson, Peg and Mac Hulslander, Carina Iezzi, Ray Kelly, Conrad Kottak, Alan LaFlamme, Megan Bishop Moore, Gerald and Lois Neher, Dot Osborne, Dan Otterbein, Steve Smith, and Ted Steegman.

My wife, Charlotte Swanson Otterbein, has assisted me on every book I have written, beginning in the summer of 1965. All subject matter is discussed with her (I am the "professor" at the breakfast table); she now types the manuscripts, since I have reverted to writing on yellow sheets with a No. 2 pencil; she edits at all stages; and she spends the royalty checks.

My son, Gere Frederick Otterbein, died unexpectedly of complications following three back surgeries. He had urged me to write this book, as well as previous books. He and I discussed this book the evening before his death.

Chapter One

Introduction
What Is the Anthropology of War?

Anthropologists study the warfare of nonliterate peoples, as well as the military organizations of all warring polities regardless of size. Some societies have become known to us archaeologically and historically and some are modern nations. Research methods include ethnographic fieldwork, archaeological excavations, historical analyses, and comparative or cross-cultural studies. The results of this research, which goes back to the mid-19th century, yielded numerous descriptions of all manner of warfare and generated a large number of theories about the causes and consequences of warfare (Otterbein 1973).

The earliest studies of warfare by anthropologists in the 19th century focused on the customs, such as marriage by capture after a raid, of a large number of cultures. E. B. Tylor, a famous British anthropologist, is credited with the first cross-cultural study. A ***cross-cultural study*** uses a sample of cultures, perhaps 20 to 100. One or more dimensions or variables, with points or categories on each, are coded for each culture. One of Tylor's variables was "marriage by capture." Tylor divided marriage by capture into three kinds: hostile capture, connubial capture, and formal capture (Tylor 1888:259). Many decades later Barbara Ayers conducted a similar cross-cultural study. She distinguished between "bride theft," which equates to Tylor's marriage by capture in which a particular woman is targeted, and "raiding for wives," in which any woman may be seized. Both practices were found to be associated with ***virilocal residence*** and general ***polygyny*** (Ayers 1974). In chapter 3 we will learn why these customs or practices are related.

In order to fill the museums of the world and to satisfy collectors of curiosities, ethnographers and other travelers collected weapons, along with many other items of material culture. Museums and private collectors in the 19th century even sought such exotic curiosities as shrunken heads, and the Jívaro Indians of the Amazon increased their raiding in order to supply more heads (Bennett Ross 1984). As I write this text, I am looking at a slashing weapon, which Higi men of Northeast Nigeria carried at night for protection (Otterbein 1968a:207). "[It] is made entirely from iron in the shape of a lower-case 'f'; the back of the neck of the 'f' is sharpened for several inches. The crossbar, which extends in the direction opposite to the cutting edge, makes it possible for a man to rest the weapon on his shoulder with the blade up while he holds it by one hand." Across the Cameroon border the same weapon is carried by the Kapsiki (van Beek 1987:56). I am surrounded by other artifacts collected in the early 1960s from the area where I conducted a study of Higi armed combat.

The term "primitive warfare" was used to describe the armed combat of nonliterate peoples until the 1960s, because native peoples were called "primitives" and their customs and weapons were considered primitive. Well-known books by reputable anthropologists from the 20th century have such titles as *Our Primitive Contemporaries* (Murdock 1934), *Primitive Warfare* (Turney-High 1994[1949]), *The Law of Primitive Man* (Hoebel 1954), and *Primitive Social Organization* (Service 1962). Today it is no longer considered appropriate to use the term "primitive." I use the terms nonliterate and native, or the actual name of the people, such as the Higi. I often use the term "armed combat" in place of "war," although armed combat can include other types of fighting with weapons, such as feuding and dueling.

New case studies of native peoples and new theories enlivened the study of warfare in the 1960s, a period I have called "The Golden Age" of the study of warfare in anthropology (Otterbein 1999b). In the early 1970s John Honigmann, the author of many textbooks and the editor of the *Handbook of Social and Cultural Anthropology* (1973), asked me to write a comprehensive review of the literature in anthropology on warfare and he assigned the title, "The Anthropology of War." I used this phrase in the last line of a review I wrote for a book on Yoruba warfare: "This book will be of concern . . . to ethnologists interested in the anthropology of war" (1966:1531).

THE BASICS

Military organizations are among the most important social institutions of any society that engages in warfare. I use the word "engage" rather than "goes to war," an expression that implies setting forth to attack another polity, because "engage" includes the notion that wars may arise because the polity is attacked. Some polities do not engage in war or have a military organization. War and *military organizations* are as important as kinship and the family, religious practices and practitioners, and the economy and modes of exchange to understanding a particular society. All of these social institutions are important to the people, and each is important as a subject of scientific research. This small textbook is an introduction to the scientific study of military organizations and warfare. Military organizations, along with governments, laws, and courts, are major components of the political system of *polities.*

Much of the research by anthropologists on primitive warfare up until the 1960s did not view war and the perpetrators of war as part of the political system of the society. One can read a section on "political organization" in an *ethnography* and find no mention of war leaders or warriors. They are likely to be mentioned in a section on "weapons." This changed in the 1960s, largely under the leadership of Andrew Vayda (1961, 1968) and his student Roy Rappaport (1968), who linked political processes to subsistence technology and warfare, an approach called Ecology. They undertook field research among the Maring, a highland New Guinea tribe. Vayda's later summary volume was titled *War in Ecological Perspective* (1976). The main chapters of the volume provided case studies of the Maori of New Zealand, the Iban of Borneo, and the Maring. My own research linked military organizations, weapons and tactics, the goals of war, and other military practices to each other and to levels of sociopolitical complexity. I did this in my ethnographic study of the Higi and in my cross-cultural studies of warfare (1968a; 1970).

Military historians, however, have often described warfare in the context of governmental functioning through time. Over a period of 20 years German military historian Hans Delbrück placed military affairs and battles in a historical context by studying wars as part of political history (1975–1985). He traced the development of European warfare from the ancient Greeks to the Napoleonic period. He argued that "the military organization is always the most basic factor in the existence of a nation" (V. 4, p. 223) and that "the entire cultural existence of peoples is determined to a high degree by their military organizations" (V. 4, p. x). Recently military historians have adopted a cultural approach to the study of war. John Lynn, in *Battle: A History of Combat and Culture* develops a "cultural model" to depict the relationships between the reality of war and how war is viewed and discussed by the participants (2003:xiv–xxii, 331–341). Cultural beliefs and values, however, have long been seen by anthropologists as influencing the way war is waged. The beliefs and values are learned, and when they pertain to warfare, this is referred to as "socialization for war." Socialization for war, thus, provides the connection between culture and warfare practices (chapter 5). Different cultures wage war in different ways.

The beginning point for any study of warfare, after identifying the nature of the *political system,* is to focus on its goals or reasons for going to war, the military organization, and the weapons and tactics employed. Military organizations are typically either nonprofessional or professional. *Nonprofessional military organizations* are composed of part-time personnel, and *professional military organizations* are composed of full-time personnel. Nonprofessional military personnel go on raids, and professional personnel go on campaigns. Each of these broad categories can be divided further—nonprofessionals into "local militias" and "Fraternal Interest Groups," and professionals into "elite warriors" and "conscripts." These four types of military organizations will be described in chapter 3. Small-scale societies, also known as uncentralized political systems, have nonprofessional military organizations, and large-scale societies, also known as centralized political systems, have professional military organizations.

The *goals of war* for uncentralized political systems are usually defense-revenge, plunder, and prestige, and for centralized political systems, the main goal is political control (Otterbein 1970:63–70). The goals increase in number with increasing political complexity. The tactics of uncentralized political systems consist of ambushes and line battles; centralized political systems use line battles and sieges (Otterbein 2004:194–202). Shock and projectile weapons can be used singly or in combination in any of these types of armed combat. Weapons and tactics are described in chapter 2, and military organizations and the reasons for going to war are discussed in chapter 2 and 3.

Two questions concern many people: Is war in our genes, and thus, is war inevitable? Is there any way to end wars, and more importantly, is there any way to prevent wars? The biological basis of war will be discussed at length in chapter 5, while the final chapter, chapter 8, will focus on the prevention and termination of war. The following highlights the direction this book takes: biology is not destiny—that is, our behavior is not solely based on our biological makeup—and the belief that war is preventable may be a major first step in creating a world with fewer wars. This belief is called *Positive Peace.*

FAMOUS WARRING PEOPLES

Of all the warring peoples studied by anthropologists two societies have come to be featured in many anthropological textbooks. The Dani of the highlands of New Guinea are immortalized in an 83-minute documentary film titled *Dead Birds.* The filmmaker, Robert Gardner, was accompanied not only by a professional film crew but also by anthropologist Karl Heider and writer Peter Matthiessen. The film has entertained and shocked generations of undergraduate students with its battle scenes of flying arrows and spears and ceremonies of young girls with their fingers cut off. The Yanomamö of South America became famous because of their portrayal as a fierce people who engage in constant warfare, and they later became infamous because their ethnographer, Napoleon Chagnon, was accused of helping to cause or facilitate their warfare. There are many other studies of warring peoples, which deserve to be read. An appendix to this book lists 40 cultures that engage in warfare, often extremely violent. These cultures are either mentioned or described in this book.

The Dani

Any study of the Dani must include a reading of Karl Heider's ethnography, *Grand Valley Dani: Peaceful Warriors* (1979, 1991, or 1997). Any edition will suffice. Also suitable is his earlier massive work *The Dugum Dani* (1970). Each describes a battle that took place after 1961, the year *Dead Birds* was filmed. The film presents only one side of Dani armed combat, what Heider calls "the ritual phase of warfare." By showing flying arrows and spears and infrequent injuries, viewers are led to conclude that Dani warfare seldom caused death and that it is highly ritualized. This is not why Heider called them "peaceful warriors." The reason is not directly provided, but the careful reader can reach an informed opinion. Many anthropologists and historians describe the warfare of native peoples as **ritual warfare** and contrast it with the deadly warfare of state societies. This characterization is wrong. Both state and nonstate societies may have high casualty rates for both combatants and noncombatants (women and children) and use ritual before, during, and after battle.

Heider calls the other side of Dani armed combat "the secular phase of warfare." In a return trip to the Dani villages in 1968, he learned that on June 4, 1966, a large Dani force attacked an enemy village at dawn, killing about 125 people and burning many of their compounds. Heider subsequently learned that sporadic outbursts of secular warfare had characterized Dani political life as long as anyone could remember. Heider's ethnographies describe both types of warfare—ritual and secular—and provide an explanation of how they are related. I will present Heider's description and analysis. I will give additional interpretation, which extends and expands Heider's treatment. It in no way rejects or disputes his conclusions, but it builds on the base he provides.

Dani political units surround themselves with watchtowers, which are manned daily by warriors armed with spears and bows and arrows. The lookouts attempt to see movement in the grasses, a sure sign that an enemy raiding party is attempting to enter their territory for the purpose of ambushing someone, even a small child or a woman. The presence of manned watchtowers is so rare among native peoples from uncentralized political systems that one can conclude that Dani warfare is extremely serious. I have often referred to these towers as unique, for I know of no other people who have developed such a sophisticated warning system. The best comparison I can think of is the radar tower network the British erected along the English Channel early in World War II, before the Battle of Britain in which the Nazi air offensive was eventually stopped by the Royal Air Force in September 1940 (Southwaite 1984:210–216).

A large number of people appear at one time on the screen in *Dead Birds*. Gatherings of warriors on the battlefield and people at ceremonies number in the hundreds. The Grand Valley has the greatest **population density** in the New Guinea highlands; the valley is about 45 kilometers long and 15 kilometers wide with a population of 50,000. Thus, in this area of about 100 square miles, one finds nearly 500 people per square mile (Heider 1970:59–60). There are about 12 polities in the valley, depending on the fortunes of war. The polities are called **alliances** by Heider, and the more stable territorial units within them are called confederacies. An alliance consists of three to five confederacies. The leaders of confederacies are rivals for the leadership of the alliance. This rivalry makes for alliance instability. In spite of the high population density the impetus for war seems to stem from alliance instability rather than from competition over land or other scarce resources.

Dani warfare as shown in the film consists of battles and raids. Battles are initiated by an important leader through a feast. A challenge is announced the next day. If accepted by a confederacy in a rival neighboring alliance, the battle begins in mid-morning. The same battlefields are used repeatedly. Raids usually take place during the morning or evening hours. A raiding party will consist of about 12 to 50 men. If they can crawl past the watchtowers without being detected, they lay an ambush. The first person coming along a path will be speared. If the raiding party is spotted, a counterambush will be set up. Raids do not occur at night.

In the period covered by the film, April 10 to October 15, 1961, eleven battles and eleven raids occurred, five by one side, six by the other. I derive the figures from three pages of field data (Heider 1979:90–92). Only two men died in battle, seven in raids. (The battles shown in the film occurred on June 10 and August 25.) Battles and raids form a system. Either can be used to avenge a death. Although descriptions of the feuding and the warfare of uncentralized political systems have often described the combat as tit-for-tat, with alternating deaths on each side, this is not what I found for the Dani. One side (A) lost six persons, the other side (B) lost three. The pattern was: A–3, B–

2, A–1, B–1, A–2. It is clear that the side seeking revenge may lose a second person and a third before they are successful in killing a member of the other side. An example of this in Western society is seen in a study of feuding in Eastern Kentucky in the 19th century. That one side is likely to end up a big loser was true in the Hatfield–McCoy feud (5 Hatfields and 9 McCoys died) (Otterbein 2000b:235; Rice 1978). Thus, in revenge-based conflicts, the score is not likely to even out over a long period of time.

This imbalance has a major effect on what can follow. The ritual phase of warfare communicates the strength of a confederacy and the solidarity of an alliance. Over time, an alliance weakens vis-à-vis other alliances. One alliance will have more deaths in battles and raids, while other alliances become larger, more aggressive, and show more skill. The weakened alliance may be the victim of a dawn attack, like the one Heider described, by two or even more alliances (1979:103–105). The losers will flee and become refuge populations among other confederacies where some may have relatives. Boundaries shift, watchtowers are abandoned and new ones erected, and fields will be planted where battlefields once were. New battlefields will be chosen. The number of alliances in the Grand Valley may decline or increase. Thus, the two phases of Dani warfare constitute a single system, a system that covers the entire Grand Valley of about 12 alliances and 50,000 people.

The Yanomamö

Vivid images of the Yanomamö appear in anthropological textbooks through stills from films produced by Napoleon Chagnon and Timothy Asch, and often through three films used in classrooms, *The Feast* (40 minutes), *The Ax Fight* (30 minutes), and *A Man Called Bee* (40 minutes). The last film shows Chagnon conducting fieldwork in Yanomamö villages. *Yanomamö: The Fierce People* was a revision of Chagnon's 1963 doctoral dissertation. It has gone through five editions. The last two editions have dropped the subtitle as Chagnon has attempted to withdraw his portrayal of the Yanomamö as "the fierce people" (1968; 1997).

The Yanomamö have become the quintessential "warring savages." Popular accounts of how warfare is ancient in the human species almost invariably include a photograph of Yanomamö warriors brandishing bows and arrows and spears—materials used by hunter-gatherer groups for subsistence. However, the Yanomamö are tropical forest *horticulturalists* who practice *slash-and-burn cultivation;* it is an error to portray them as representative of early humans who for hundreds of thousands of years were hunters and gatherers, not horticulturists.

Yanomamö villages, scattered throughout the tropical forests of the Amazon, are polities that number about 40 to 240 people. I have calculated the population density as .25 persons per square mile, assuming 10,000 people in 40,000 square miles (1977:46). Each Yanomamö village is surrounded by a tall-palisaded wall, which prevents surprise attacks by raiders. Raiding parties make their kills by lying in ambush along trails, especially those outside enemy villages. Each village has a **headman** with limited power; he leads but cannot coerce. On rare occasions, there are co-headmen. Each village has two or more intermarrying "local descent groups" composed of patrilineally related males. Their wives come from the other descent group. The Yanomamö are **patrilocal**, **patrilineal**, and **polygynous**. When villages become large they are likely to fission, forming a new village, made up of sections of each local descent group.

Villages form alliances. Unlike the Dani, these alliances are not polities but are true alliances between polities. The headmen of two villages will make a secret arrangement to attack a third village in order to seek revenge by killing enemies in the targeted village and to steal women for wives. (E. B. Tylor would have classified this as "hostile capture.") Since one of the two villages may have a trading relationship with the targeted village, it is possible for a "treacherous feast" to be arranged. Both men and women from the targeted village are invited to a feast. The village that is allied with the host village will surround that village while the festivities are taking place. At a chosen moment the hosts attack their guests. The women are seized while the men are speared. Those who try to flee the village are attacked by warriors from the allied village. The population of a village can be nearly wiped out in this manner. With about 125 Yanomamö villages, with many villages forming alliances, and with many villages fissioning, the Yanomamö region forms a warfare system in flux, just as did the Dani in the Grand Valley.

In addition to raids and treacherous feasts, Chagnon describes three other types of violence that occur between Yanomamö villages: (1) Chest-pounding duels, in which two men take turns, each hitting the other in his chest with a bare fist. (2) Club fights, in which groups of men swing long poles at each other with the intention of knocking the person down or unconscious. Club fights may escalate to using machetes or axes, as shown in *The Ax Fight* film. Club fights can also occur within a village, with the outcome being village fissioning. (3) Spear fights, which are arranged battles, rarely take place. (See chapter 4 for further analysis.)

Chagnon's theorizing in the 1980s led him to the argument that fighting over women resulted in successful warriors having more wives and children than men who had not killed in combat (1988). Thus, successful raiding and reproductive success go hand-in-hand. Never stated by Chagnon, but I believe inferred by others, was the idea that a gene for aggression was being increased within the Yanomamö population. (Chagnon told me he did not know because he did not now how long the Yanomamö had been warring.) Aggression is not warfare. Warriors' children might be more likely to become warriors through socialization. Chagnon's finding has received severe scrutiny by Brian Ferguson (1995:361–362), Douglas Fry (2006:288–305), and others (reviewed in Fry 2006:184–199). For methodological reasons and other reasons, it has been almost universally concluded that successful warriors do not have more children than other men.

Long before this controversy erupted in the late 1980s, I had realized that males who go to war may not survive and therefore they may not produce offspring. I would joke in class that the man who stayed home got to impregnate the women, but overseas the soldier might spread his genes. Then, I became more serious. I would talk about fighter pilots. Flying a fighter plane, in or out of combat, is probably the most dangerous occupation in the world, exceeding Formula I racing and bull fighting. I know a pilot who was married and had four or five children at the time, and was not permitted to go into combat for this reason. A distant cousin of mine, a naval pilot, crashed and was killed before he had a chance to marry. A third example was my friend and neighbor, Dan Sheedy, who fought in the Battle of Midway, June 4, 1942, and survived to marry and have five children. His role in the battle is described in accounts of the unsuccessful torpedo bomber attack on the Japanese aircraft carriers. Later in the battle, U.S. dive bombers sank the four Japanese carriers. Dan piloted a Grumman 4F4 Wildcat from the deck of the aircraft carrier *USS Yorktown*. Attacked by a Japanese Zero (a lightweight fighter plane) and wounded in the right leg and shoulder, Dan was nevertheless able to drive off a Zero that was attacking a squadron mate. Attacked again by Zeros while flying just above the ocean, he was able to turn and, in a head-on pass, with guns blazing, send the Zero into the sea. He was forced to crash land on the *USS Hornet*, since without a working compass and instruments, he could only estimate where the three United States carriers were located. In February 2005, as Dan was being wheeled on a gurney to an ambulance he told his neighbor Fred to "tell those guys if it weren't for me they would all be eating sushi." Dan did not return from the hospital. The point of my story is that men who survive combat may have children. This, of course, pertains to the Yanomamö research. Those who die young do not. No conclusion can be reached that war leads to individual reproductive success.

I am on record as arguing that Yanomamö warfare "can largely be explained by *Fraternal Interest Group theory*. . . . The Yanomamö alliance system is also an important factor in intensifying warfare. . . . A third factor contributing to the high levels of Yanomamö violence is probably the brain damage that is likely to result from the staff fights that mutilate the craniums of virtually every adult Yanomamö male" (Otterbein 1985a:xvi; emphasis added).

Three charges have been lodged against Chagnon's characterization of the Yanomamö as "the fierce people." I believe these charges cannot be substantiated. The first charge is that he misrepresented ethnographic reality; other anthropologists describe many Yanomamö villages as peaceful or nonviolent. Sometimes Yanomamö violence is classified as feuding, not warfare (Sponsel 1998). Numerous anthropologists consider feuding to be judicial in that the self-help taken by a kinship group is considered legal (Fry 2006:108–113). Second, some scholars assert that Chagnon engaged in practices that engendered fighting, such as giving the villagers with whom he lived steel tools, including machetes and axes. The questions he asked brought forth aggressive attitudes and responses, and the films that he and Timothy Asch made were staged and constructed. The third challenge to Chagnon's claim is that the violence Chagnon reported was a reaction on the part of the Yanomamö to culture contact, which began in the 17th century. Slave raids over two centuries, a rubber boom of the late 19th century, and the resulting migrations and wars drove the Yanomamö into the highlands. The establishment of mission stations in the lowlands in the 1950s brought the Yanomamö out of the hills, eager to obtain steel tools; villages fought to monopolize the trade (Ferguson 1992:202). For a balanced review see Bruce Bower (2001).

What is true is that Chagnon selected a people with a background in warfare and picked out the most violent villages in which to live. After all, warfare was the focus of his research, and thus, the bulk of his field records contain information on warfare. When he wrote his dissertation (I have a copy) and later revised it for publication, it continued to focus on warfare and other kinds of violence. I did the same in my research with the Higi of the Mandara Mountains. I selected a warring people, focused on a village that had recently gone to war, focused on warfare, and wrote an article that focused on warfare and other types of armed combat and how they were related to the sociopolitical organization. I believe that many of us who have studied warfare, feuding, and related topics have been attacked at one time or another either for being advocates for war or for helping those who wish to wage war. All those whom I know who study war are antiwar.

The above discussion of researchers of and research on war and the peoples labeled as violent or prone to war illustrates the complexity of the topic and how it is interpreted. It is my hope that the chapters that follow provide clear explanations for the perspectives of what constitutes war, why it exists, how it is carried out, and how it changes (or not) over time, among people, and within different contexts. My hope is that the reader will have a better understanding from which to draw his or her own conclusions and evaluations.

Chapter Two

Weapons and Tactics

Weapons and tactics go hand in hand. A tool or artifact is a weapon only when it is used to make contact with a target, animal, or person; this can be done by, for example, firing a rifle, releasing an arrow from a bow, or launching a dart from a spear thrower or atlatl. In the anthropology of war these acts could represent "shoot-on-sight" tactics—an expression used by Raymond Kelly in his study of Andaman Islander warfare. Pygmies on the Andaman Islands in the Sea of Bengal, north of the Indian Ocean, are hidden when they shoot trespassers, using poisoned arrows (Kelly 2000). This tactic is an ambush. The employment of weapons in tactics usually results in injury or death to the person attacked and sometimes to the attacker if the object of the attack has and can use his or her weapon. Thus, the last section of this chapter will discuss casualties in combat and in its aftermath.

WEAPONS

The first dictionary definition of a **weapon** is "any instrument used in combat." Years ago I adopted the term **armed combat** and defined it as "fighting with weapons." I have identified five elements defining a weapon: (1) it is an object—natural or man-made; that is, it is a tool. (2) It is used to make contact with something—a target, an animal, or a person. (3) It can be hand held (shock) or thrown (projectile). (4) Accuracy is sought by the user. (5) Motives for its use can vary from scoring to injuring to killing.

Weapons can be used in close combat or projected from a distance. Primatologists use the terms "stick-club" and "missile-throw" to describe "anti-predatory patterns of weapon use" by wild chimpanzees (McGrew 1992:180–181). Military historians have long used the terms shock and projectile (missile) (Oman 1960[1885]). A shock weapon is kept in the hand during combat. If it is thrown it is a projectile. So, a spear used for thrusting is a shock weapon and a spear thrown is a missile. A throwing spear can be called a javelin, and if its flight is assisted by an atlatl, it is likely to be called a dart (Dickson 1985).

Shock Weapons

The club is the basic shock weapon. It was also probably the first weapon. When an early member of the human line picked up a stick and hit a small animal or a rival who was trying to seize his companion, that stick became a club, a weapon. Over the millennia, the club has often become elaborate, taking many forms. The head of the club can be a round or oblong weight that is a stone or metal device. Nonliterate peoples at all levels of sociopolitical complexity used clubs as did many literate peoples before the invention and introduction of firearms. Maces were used by ancient Mesopotamians and Medieval knights. Plains Indians such as the Sioux and Cheyenne used war clubs. These are the Native Americans who defeated General George Custer at the Battle of the Little Big Horn. They also had projectile weapons—Henry and Winchester repeating rifles and, as the battle progressed, they picked up Remington single shot carbines dropped by wounded and fallen soldiers (Scott and Fox 1987). Archaeologist Richard Fox has synthesized the battlefield finds with Indian accounts of the battle (1993), while historian Gregory F. Michno focuses on the Sioux and Cheyenne narratives of combat (1997).

Knives and swords are also shock weapons since they are held in the hand and put into use by stabbing or slashing. The Higi weapon on my wall can be used for both slashing and striking with a downward motion. I do not know when a knife becomes a sword—at 10″, 12″, 14″? Materials for knife and sword blades can be stone, copper, bronze, iron, steel, and today, a synthetic product. Some modern weapons may be called "shock" weapons because they deliver a shock effect, "shock and awe."

Projectile Weapons

A projectile weapon is thrown, launched, or shot by a combatant or machine designed to propel the missile. Hand grenades, exploding artillery shells, and bombs dropped from airplanes are classified as projectile weapons. The range can be close, just below the airplane, or half the world away with an ICBM (intercontinental ballistic missile) used to send the explosive to its target. In range, striking power, and rapid firing, the compound bow used by the armies of the ancient Near East, from Egypt to Sumer to Mongolia, was the most effective weapon invented until the development of firearms. Figure 2.1 gives a graphic comparison of the effective range of weapons used by prehistoric, nonliterate, and early historic peoples.

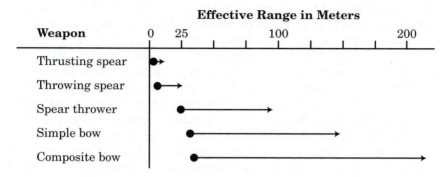

Source: Otterbein 2004:66.

Figure 2.1   Effective Range of Weapons

Rarely can a missile be stopped in flight, but shields and/or body armor, including helmets, can be used to deflect or block incoming projectiles. Often shields and body armor are highly decorated; the decoration itself may be intended to help deflect incoming missiles. Chariots, armored vehicles, or tanks may provide protection from missiles. Combatants can erect field fortifications or take up defensive positions behind settlement walls to protect themselves (Otterbein 1970:49–62).

TACTICS

Tactics are the "science" of placing and maneuvering military organizations for combat. Leaders develop a plan of attack (ambush, line, or siege operation) or a plan of defense (fortifications). Tactics range from simple to complex. The most simple is the shoot-on-sight tactic, usually from ambush, and the most complex would be a coordinated attack by land and sea forces. The use of air forces in the 20th century made for even greater complexity. Probably every war leader from prehistoric times to the present has made the distinction between ambushes and lines (Otterbein 1970). An ambush will be tried first, and, if it is not successful, a line is likely to be formed. If the enemy are behind their walls they will be challenged to come forth and engage in combat. If they do not accept the challenge, the forces on the offensive may set up a siege. Siege operations can be considered a tactic, but they occur only if the enemy's settlement or city is fortified (1970).

Ambushes

Ambushes are nearly universal, either laying a trap or surrounding a village or campsite. They are used in warfare, feuding, homicides, and political assassinations. An ambush is typically employed by a group, sometimes by a single attacker or marksman. The Dani and Yanomamö used both forms of ambush. In the first type of ambush the warrior(s) hides along a trail, path, or road and permits the enemy to walk, ride, or drive into the trap. In the second type the warrior(s) attacks when people leave their settlement or, at a given signal, the warrior(s) attacks from all sides. The Yanomamö attack people as they leave a settlement to urinate or to flee from a treacherous feast, while the Dani attack a settlement at dawn while people are sleeping. Successful attackers see the killings as showing great acumen or skill. The victims see the attack as a cowardly act by a loathsome people, as did U.S. citizens after the attacks on Pearl Harbor (December 7, 1941) and on the Twin Towers (September 11, 2001).

In a study of Kentucky feuding, I found both types of ambushes frequently used (2000b). My study included five different sets of feuding families. In the 14 killings in the Hatfield–McCoy feud, three occurred in an ambush and two in a home attack. For all five feuds analyzed there were 64 deaths; 28 were in ambushes (44%) and 9 in home attacks (14%). In second place were gun fights, a chance encounter between feudists who typically carried arms at all times, amounting to 10 deaths (16%).

In warfare since the rifled barrel was developed, snipers have taken their toll on enemy officers. Snipers even hunt each other. A fascinating firsthand account is given by Carlos Hathcock, in *Marine Sniper: 93 Confirmed Kills* and *Silent Warrior* (Henderson 1986, 2000). In the jungles of Vietnam Hathcock faced a North Vietnamese sniper. Each man was equipped with a scoped bolt-action rifle, the choice of many snipers, because it allows the shooter to easily target the victim; the spent cartridge remains in the gun, not outside where it can be discovered as evidence; and the gun can be easily fired from the prone position. Both men fired nearly simultaneously, but Carlos fired first. When he recovered the rifle from the body of the Vietnamese, he found the optical lens of the scope had shattered and the bullet had gone into his eye, through his brain, and out the back of his head—his enemy had the crosshairs directly aligned on Hathcock (Henderson 2000:202).

Lines

Line battles may be either arranged or the result of chance encounter. In an arranged battle, leaders or representatives of both sides select the time and place for the combat. The Dani's traditional battlefields were named and repeatedly used. So were those of the Higi (see Heider [1991:102] and Otterbein [1968a:199] for maps of locations of battlefields). Arranged battles are not irrational. If both sides are willing to engage in armed combat, it seems reasonable to select a field of battle satisfactory to both. Perhaps the operative words are "satisfactory to both"; history is replete with cases of one side drawing up in a defensible location, such as a hilltop or hillside, and the other side refusing to attack. Another possibility is that two armies may campaign all season long and not find each other. In desert areas of the Middle East or North Africa this would be possible, but not in a densely populated region like the Grand Valley or Mandara Mountains. Encounter battles occur when two war parties or armies meet in the field. One side may not wish to fight but is forced to do so. Such a side may attempt to display great strength but will leave the field in an orderly fashion as soon as it can. An encounter battle may arise if an ambush fails to be successful for those who laid the trap.

Lines may consist of specific units—each combatant belongs to a grouping of fighters (a unit) who have trained together. Units are likely to be roughly the same size. They arrange themselves on the battlefield in an order based on tradition or in an order designated by the commander. The men of a unit may have grown up together in one village. They may also be members of the same age grade, a group that went through initiation together. In South Africa, Dingiswayo (c. 1780–1817) was the leader of the Mtetwa and Zulu and organized young men according to age grade regiments that went into battle together. Sometimes a particular position in the line is the place of honor. Often it is on the right. Officers may position themselves in front of or behind their units. The leading officer, possibly the supreme commander or monarch, may station himself behind the line, where he can see the action as it unfolds (Keegan 1987).

In polities where kinsmen fight together, lines may lack specific units. Indeed, this is the common practice in uncentralized political systems. There is no training together as combatants, although the senior male in the group is likely to assume command, as he does when he and his kinsmen have hunted together. Men will charge and retreat when they please. This is the type of line that we see in *Dead Birds*. A warrior takes a position in line wherever he wants. For the Higi, a man makes sure he is not opposite an in-law. His wife may have come from the group he is fighting and he dare not injure or kill one of her relatives (Otterbein 1968a:211).

Another type of line is made up of units that differ. The soldiers in one unit may be equipped with shock weapons, those in another with projectile weapons. Some units may be cavalry, others chariot corps. Sometimes units consist of heavily armored infantry. The soldiers wear helmets and body armor, and each carries a shield and a shock weapon such as a club/sword and a thrusting spear or pike. Light infantry, on the other hand, are equipped so they can run, and they carry light spears or bows and arrows. What is key to this type of line organization is that the different types of units coordinate with each other. Cavalry units may occupy the ends of the line; light infantry may proceed forward of the heavy infantry; archery units may stand behind the infantry. A Higi shield bearer would stand in front of a Higi archer, the two forming a single combat team that moved together, but some archers preferred to dodge arrows (Otterbein1968a). Lines with differing units are typically found in centralized political systems, some equipped with missiles, others with shock weapons. And cavalry and infantry units may also be found in centralized systems as well.

Not until the development of complex, centralized political systems do we find tactics that use in combination all three of the following: artillery or catapults, chariots or cavalry, and light or heavy infantry. This is the *integrated tactical force* loved by military historians (Ferrill 1985:150). It is the quintessential "modern" military that has been used from the time of Alexander the Great to the present. Its use was perfected by each "great captain" or military genius, such as Frederick the Great or Napoleon Bonaparte. The U.S. Army refers to this as a "combined arms team," which is defined as "application of several arms, such as infantry, armor, artillery, and aviation" (Department of the Army 1962). The only thing the ancients did not have was aviation, but it was envisioned by geniuses such as Leonardo da Vinci. He designed a flying machine, a helicopter or aerial screw; a parachute; and, for land warfare, a scaling ladder; a triple-tier machine gun; a multiple-barreled cannon; and a military tank shaped like an inverted funnel.

Siege Operations and Fortifications

Siege operations are the means to destroy, capture, or prevent the escape of an enemy in a fortified location, as well as the material aids such as ladders, battering rams, or shovels that are used in the siege. The site, most likely a city, will be surrounded. If the enemy does not surrender, techniques will be used to go over, under, or through the natural barriers and constructions that the defenders have used and erected to prevent a successful enemy attack. Siege techniques do not develop unless there are fortifications to surmount (Otterbein 1970:58–63). Fortifications, usually walls, will be erected when there is a threat from an enemy. However, walls may be built for reasons other than defense; moreover it is possible that a wall could be built for defense, yet never see action. Some of the reasons for wall construction are to keep in domestic animals and children and to keep out wild animals and thieves from other villages. Other reasons include protecting the site from floodwaters, dust storms, and the winds of tropical storms, and to provide a raised roadway. Walls built for any of these reasons may later be used for defense (Otterbein 2004:188–194). Just because an archaeologist finds a wall does not necessarily mean warfare occurred in the region. When I see a wall, before I conclude that a battle was fought, I look for arrowheads lodged in the wall or other evidence.

A moat or ditch constructed in front of a wide, raised wall, with towers erected on it, in front or behind, is a basic scheme for defense.

The towers can serve as lookout posts and they can have platforms from which defenders can fire projectiles. An important point is that defenses have to be *actively* defended; that is, they can repel attackers only if there are warriors, even women and old men, who use the wall for personal protection and for advantageous firing positions.

There are ways other than walls to fortify or protect a settlement. The placement of a settlement along a river, lake, or ocean means an attack can come from only three directions, unless the enemy has boats. The placement of a town on a peninsula blocks attack from three directions, but it may also prevent the residents from fleeing by land. In the American Revolution, American General Benjamin Lincoln and his army were trapped in Charleston, South Carolina, a city on a peninsula (they surrendered May 12, 1780), and English General Lord Charles Cornwallis and his army were trapped at Yorktown when the French fleet defeated the British fleet that arrived to rescue them (they surrendered October 20, 1781). Probably the best location for a city is on a large hill, provided there is a source of drinking water that cannot be cut off by the attackers. The enemy would need a large army in order to surround the hill and prevent escape.

Walled settlements can have other means of defense. The wall may include locations prepared for the use of defenders. People can dig pits and place sharpened stakes in the center, covering the stakes so they become the equivalent of land mines. Like mines, they can kill children of the defenders who stray from safe areas. Hedgerows, another means of defense, can block the forward movement of attackers and prevent them from seeing their destination. The Yanomamö lived in palisaded villages and hid the paths through the jungles that led to the villages. The Higi built their stone-walled round compounds on hillsides. Compounds in a dispersed settlement were connected by a maze of cactus hedges, one on each side of a path. New hedge walls were constantly being constructed. Certainly they had a defense in depth that prevented slave raiders on horseback from getting into their compounds (Otterbein 1968a:198).

A siege operation is a special kind of line. It need not be straight, for it may need to encircle a hill. It needs to be designed to resist attack by the besieged and to prevent them from breaking through to escape. The attacking army needs to have a steady flow of supplies. Siege equipment needs to be at the ready and to be kept repaired. Typical siege equipment includes battering rams, sometimes housed in covered sheds to protect the operators of the ram from projectiles dropped on them. Siege towers that can be moved forward, often on wheels, have platforms that are higher than the walls, thus permitting the attackers to fire down on the defenders on the walls. More simple are shovels used to tunnel under the walls and tall ladders that can be leaned against the walls and climbed. In the technologically advanced societies of the ancient world, catapults could launch large rocks and boulders that might shatter gates, towers, and even walls. With the development of cannons in the 1400s, walls surrounding cities became useless as defenses. Will Mariner and crew members used fixed mounted cannonades to assist Tonga chief Finan to slaughter the forces of his enemy within a stockade constructed of wickerwork (Michener and Day 1958:267–269). Chief Finan became King Finan II.

BASIC PATTERNS OF COMBAT

After studying the **warfare systems** of many nonliterate societies at all levels of sociopolitical complexity, I identified two basic patterns of combat. Although warfare systems are unique, I find that most uncentralized political systems use ambushes and lines in combination and that many centralized political systems, most of whom have fortified towns, use lines and siege operations in combination. The lines of each differ. As described above, some lines are informal in that a warrior can take any place on the battle line; uncentralized political systems use this type of line. Other lines are formal with similar units occupying fixed positions along the line; centralized political systems use this type of line. Integrated tactical systems are a further development of the formal line.

Ambushes and Lines

The major features of this pattern have been described for the Dani. The strength of each side was tested in a line battle, which resulted in few casualties and deaths, and in ambushes, which resulted in many more deaths. Ambushes and lines occurred over time in no regular sequence. One side, however, suffered more deaths. "Line battles appear to be a testing of strength, whereas the ambush is the tactic chosen to inflict great casualties or to destroy an enemy village" (Otterbein 2004:202). Dani villages were destroyed by ambushes. In other societies the group that has been weakened may flee.

Another case study of the basic pattern of ambushes and lines comes from a hunting and gathering society in the Northern Territory of Australia, the Tiwi, studied by ethnographers C. W. M. Hart and Arnold Pilling (1960). The Tiwi live on Melville Island off the northwest coast of Arnhem Land. They engaged in arranged spear-throwing battles, which resulted in low casualties, and sneak attack raids, which occurred at night and resulted in high casualties. The line battles occurred before and after British contact. A 1928 line battle is described by the ethnographers. It is a classic "ritual battle." Before European contact, night raids occurred. Pilling wrote in *The Tiwi* that his cases listed 54 deaths and 19 injuries from sneak attacks (1988).

"Those killed represented over ten percent of all males in that age category" (Pilling 1968:158). It seems clear to me that the line battles tested strength, while the night raids inflicted the heavy casualties (Otterbein 2004:83).

Beside skeletal trauma and stone points in bone, the only seeming evidence for late *Paleolithic* warfare are paintings on the walls of caves (Otterbein 2004). In Europe, the earliest are crude figures with missiles protruding from the body. Many of these figures are shown in *The Origins of War* by a prehistorian and a medical doctor, Jean Guilaine and Jean Zammit (2005). Since the figures are sometimes superimposed over game animals, Raymond Kelly has offered the interpretation that a trespasser from another group has been slain and that the paintings "memorialize spontaneous confrontations over game resources in which the social group of the painters prevailed" (2000:153–154). Later paintings in Spain show a firing line of 10 archers and a victim on the ground pierced with 10 arrows. A number of paintings dating from 10,000 to 4,000 years ago in northern Australia show figures confronting each other with boomerangs and barbed spears. Spears are shown plunged into fallen figures. (Remember, this is near the area where the Tiwi live.) The more recent paintings from this period show battle lines with weapons flying overhead and fallen dead (Tacon and Chippendale 1994). These are probably the ancestors of the Tiwi.

It is conjectural, but I see in the cave paintings both ambushes and line battles that formed the basic pattern of warfare for the Tiwi. Archaeologists refer to this inferential technique as a "bridging argument." Specifically, it is called the "direct historical approach." Ethnographic and historical data from the same area are used to provide meaning and interpretation to the archaeological materials found (Otterbein 2004). I also see for Europe the same basic pattern. If I am correct, the basic pattern of ambushes and lines goes back perhaps as early as 40,000 years ago and possibly back to the origin of war (see chapters 5 and 6).

Lines and Siege Operations

The line battles that combine with siege operations require a high degree of organization and control that develop only in centralized political systems. For the pattern of lines and siege operations to work, officers and commanders must be able to demand obedience to orders. Further, these politically complex polities often use diplomacy in conjunction with line battles and siege operations in pursuing their military goals.

The high degree of control required to conduct successful line battles and siege operations I have called a "coercive command structure" (Otterbein 1997:265). Military personnel are best referred to as "soldiers," who are expected to obey orders, rather than "warriors," who are encouraged to act violently. I find it regrettable that the U.S. Defense Department calls U.S. combat personnel warriors; many acts of violence in war can be considered war crimes.

An overgeneralized account of how states, politically complex polities, engage in diplomacy follows. One state attempts to force another state to give up land, resources, or anything else that it may wish, such as works of art or women. (States will be discussed in more detail in chapter 3.) Negotiations will take place and the state that views itself as militarily stronger will use **coercive diplomacy,** namely, threaten the other side with a military attack if it does not oblige. Political scientist Seymour Brown describes how coercive diplomacy can fail; that is, war begins. He gives the following reasons: (1) persisting with non-negotiable demands, (2) calling bluffs when threats are real, and (3) getting into a "game of chicken" (1987:79–83). Thomas C. Schelling in *Arms and Influence* (1966) provides an exhaustive treatment of coercive diplomacy, and he provides examples from the Comanches to the Mongols under Genghis Khan.

If negotiations break off, the aggressor will invade with an army that it deems sufficient for success. Surprise is probably not possible. The targeted polity, if it wishes to defend itself, has one of two choices: it can fight or remain behind the walls of it cities. An encounter or arranged battle may take place. In a frontal attack by both sides, each will attempt to force the other side to retreat or panic. Battles may last only a short period of time. If the defenders win they will usually pursue and capture and kill as any of the enemy as they can. If they lose, they will retreat to the nearest fortified city. The victors are likely to issue an ultimatum. Surrender and make the concessions we have demanded, such as becoming a vassal state or slaves, or we will storm your city and kill everyone, including the women and children. Historically this has been no idle threat. For examples, see the Hebrew Bible and the *Art of War in Biblical Lands* by Yigail Yadin (1963). For longer general accounts see my earlier writings (1997:262–266; 2004:187–189).

CASUALTIES

Regardless of which basic pattern of combat is followed by the people and polities of a region, casualties may be high. Each basic pattern has evolved in a manner that makes possible the killing of combatants and noncombatants in large numbers. Atomic bombs are not necessary.

Casualties in Combat

Although battle statistics are rare, it is possible to estimate from many ethnographies which types of armed combat inflicted the most

deaths. Heider's ethnography of the Dani is exceptional. The line battles and small-scale ambushes during five months resulted in nine
deaths. Five years later 125 people were killed in a dawn attack on a
village. Chagnon's ethnography of the Yanomamö makes it clear that
frequently someone who is alone is killed in an ambush. The desire to
kill more men at once and steal women led to the development of the
"treacherous feast." It appears for both the Dani and the Yanomamö
that the large-scale ambushes of people in villages were the result of
each side trying to break a stalemate. Thus, when warriors engaged in
line battles, using primarily projectile weapons, casualties seemed low;
these warring societies, nevertheless, devised other tactics and used
shock weapons, which result in high casualties. This pattern probably
occurs in many uncentralized political systems.

A state, previously a tribe and chiefdom, for which battle statistics are available is the Zulu of South Africa. One historical account of
the life of Shaka, a legendary commander and despot, provides figures
for six battles stretching over 16 years (Ritter 1957). Anthropologist
Max Gluckman, the leading ethnographer of the Zulu, accepts E. A.
Ritter's account and provides a lengthy analysis of Shaka's rise to
power (1974). In the first two battles that took place in 1810 and 1813,
a period in which the Zulu were a tribe, the average numbers of combatants killed on both sides were low, 35 and 325 (3% and 15% of the
total combatants on both sides). With the development of political centralization, age-grade regiments, reliance on shock weapons (a short
stabbing spear with an iron blade), and much larger armies, the average number of combatants killed skyrocketed: 300, 4,550, and 11,000
(44%, 60%, 80%). The battles took place in 1816, 1818, and 1819.

By 1820, large kingdoms had developed with huge armies. Forty
thousand were in the Zulu army. In the 1826 battle in which there were
English observers, an average of 12,000 were killed on both sides (42%).
Shaka was assassinated by his three half-brothers on September 22,
1828 (Morris 1965:120). Throughout the 19th century the Zulu continued to fight all comers, including the British army (Edgerton 1988). In
the end, the Zulu were no match for the British army, which had a
higher degree of military sophistication than the Zulu had. The homesteads of the Zulu and other Nguni-speaking cattle peoples were
defended only by surrounding fences that kept in the cattle. Thus, there
was no need for siege operations to arise (Otterbein 1964a). The above
study of Zulu battles gave rise to a testable **hypothesis:** the higher the
degree of military sophistication, the higher the casualty rates.
Another study, *The Evolution of War*, confirmed this (1970:81–84).

Treatment of Captured Enemies

A cross-cultural study of the killing of captured enemies provides
several generalizations that are consistent with the discussions of the

basic pattern of combat and casualties in combat (Otterbein 2000c).
Nearly all societies kill captured enemy warriors, and in only approximately one-third of the societies were warriors always spared or sometimes spared. The polities most likely to kill captured warriors or
soldiers were centralized political systems. The only centralized polities that did not were states that had the same governmental structure
for over 200 years. Those centralized political systems that killed soldiers also killed women and children. Early states, newly formed states
that have had the same government for less than 200 years, are typically despotic. They torture and kill their own population in order to
terrorize them into submission (Otterbein 1986). These despotic states
also kill men, women, and children in other polities. Early or despotic
states will be described further in chapter 3.

The vast majority of uncentralized political systems spared
women and children. These societies often took women as mates and
adopted their children. Many uncentralized political systems are polygynous, a cultural practice that makes it possible for adult males to
acquire additional wives. Recall that in the opening section of this book,
"raiding for wives" was a feature of societies that were patrilocal and
polygynous. Some centralized political systems also took women for
mates. More than half of the centralized political systems took captives
(men, women, and children) as slaves or sold the captives into slavery.
More than one-fourth of the uncentralized polities took captives and
sold them to members of centralized political systems. Sometimes captives, particularly valiant warriors, were sought for human sacrifice. In
a small cross-cultural study of human sacrifice, I found that despotic
states commonly performed human sacrifice in public and with great
ceremony (2004).

Chapter Three

Military Organizations

Military organizations and polities are closely linked. Each polity and its military organization, if it has one, is unique; no two pairings are alike. Because there are so many, readers will gain a better understanding by looking at a limited number of types or classifications of polities and the military organizations typically paired with each type. Similar political communities have similar military organizations. First, we will look at the types of political organizations.

POLITIES AND THEIR MILITARY ORGANIZATIONS

The four types of polities I previously identified and used as the bases of discussion in this and my other books differ substantially from Elman Service's evolutionary scheme or typology—bands, tribes, chiefdoms, states. I have in recent years rejected his four types as a tool for classifying societies. Bands, in Service's (1962) scheme, were hunter-gatherers who became tribes when they developed unilineal descent groups, called sodalities by Service. The addition of economic **redistribution** in the hand of a chief produced a chiefdom. When a government with the legitimate use of force developed, a state appeared. In my cross-cultural study of warfare practices I used these four types to sort my sample of 50 cultures (1970). Tribes and states occurred most frequently. I grouped the bands and tribes, and called them "uncentralized political systems," and I grouped the chiefdoms and states, and called them "centralized political systems." Prior to the 1960s, the typology most frequently used for research on polities was state and stateless societies (Fortes and Evans-Pritchard 1940). Some anthropologists have adopted my convention of grouping chiefdoms with states. We have no known example of any society beginning as a band and evolv-

23

ing into a state. A jump from one stage to the next is all we observe. The Zulu people of South Africa developed from tribes to chiefdom to state—three linked stages. I have described the changes in Zulu military organizations and warfare practices that accompanied each shift (1964a). I am still looking for another example.

I have retained the distinction of uncentralized and centralized political systems, but I now see a fuzzy line between these two types. The Higi fall on this line, as do other societies, such as the Cheyenne of the Great Plains and the Iroquois of the Northeast. All three societies are said to have *chiefs,* but they do not fit Service's definition for a chiefdom. They have no economic redistribution—Service's criterion for a chiefdom. Each of these groups has a quite different sociopolitical system; their military organizations scarcely resemble each other.

I have divided the polities that engage in war into four types. Two of these types are uncentralized political systems, and two are centralized political systems. Type A and Type B Societies are uncentralized political systems, while Early States and Mature States are centralized political systems. The type A and B polities do not evolve into each other, and Early States might never become Mature States, but could under the right conditions. The polities and the military organizations pair as follows:

Type A Society—Fraternal Interest Groups

Type B Society—Village Militia

Early State—Elite Warriors

Mature State—Conscripts

Type A Societies—Fraternal Interest Groups

A Type A Society has a Type A personality.

Type A Behavior Pattern is an action-emotion complex that can be observed in any person who is *aggressively* involved in a *chronic incessant* struggle to achieve more and more in less and less time, and if required to do so, against the opposing efforts of other things or other persons. (Friedman and Rosenman 1974:84, italics in original).

Type A Societies are aggressive and involved in a chronic, incessant struggle to achieve or acquire more and more at the expense of another society. The "more and more" may be wives, livestock, land, or any material resource. The males of such a society seem to have insatiable wants; they will hate and fight any society that attempts to acquire the same resources (Otterbein 1999a).

Competition also takes place within the Type A Society; it is filled with violent strife. Men compete for the headmanship, sometimes killing rivals or causing them to depart. Fights and homicides occur. Wife stealing and rape are frequent. Feuds occur between kinship groups.

Headmen may be assassinated and villages may fission. Men carry weapons developed for self-defense outside the home. They develop specialized weapons for use in raids, ambushes, and line battles, which they also may regularly carry. In some cases, men may wear body armor when they are away from the village. Andaman Islanders wore body armor while hunting in order to give themselves some protection from arrows fired by ambushers (Kelly 2000). In some countries, such as the United States in the 21st century, both men and women carry canes, knives, or handguns for self-defense. However, carrying weapons is not unique to Type A Societies.

Type A Societies have existed through the millennia and on all continents from their earliest settlement by Homo sapiens. Mode of subsistence does not seem to matter. Technologies have varied greatly, from horticulture, herding, and big-game hunting, to fishing. Generally, Type A Societies occupy villages or settlements of *dispersed homesteads.*

Each village or settlement is a separate polity with a headman who is often a leader of raiding parties. His leadership is based largely on consensus of the males in the village, particularly the older males who may form a council. In a Type A Society men compete and try to dominate each other. A form of "coercive diplomacy" characterizes the relationship between many males. The most successful practitioner of coercive diplomacy will probably be the man who becomes the headman (Otterbein 2004:77–81). Societies with this political structure are said to be egalitarian (Woodburn 1982; Boehm 1999) although some are more egalitarian than others. I believe, though, that the Type B Societies described next are more egalitarian than Type A Societies.

The support of male kin is critical for the headman. Indeed, competition between separate groups of related males, called *Fraternal Interest Groups,* leads to the violence characterizing Type A Societies. There are many reasons why Fraternal Interest Groups arise. One reason is that cultural practices keep related males in their natal village— the village of their birth. Men can cooperate with their sons, brothers, and brothers' sons. Even when men marry, virilocal or *patrilocal residence* keeps them near their families of origin, because their wives typically come from other villages. The practice of marrying a woman who is not a member of one's descent group is called *exogamy.* Even if men obtained their wives from within their village, Fraternal Interest Groups could exist if they continued to live near or with the man's relatives. The practice of polygyny also produces Fraternal Interest Groups. *Polygyny* is compatible with patrilocal residence, since a man can be married to several women who reside with him or in houses within his compound. The head of a polygynous family may have many sons, who are brothers or half brothers. In polygynous societies, raiding for females who can become wives creates constant warfare between neighboring polities. In a review of Fraternal Interest Group theory two primatologists argue that societies with subsistence based on pastoralism, intensive horticulture, and agriculture tend to have strong Fraternal Interest Groups (Rodseth and Wrangham 2004:395–398). They point out that Raymond Kelly has argued that "when food is stockpiled, it inevitably becomes a key military objective of raiding parties" (2000:68). Furthermore, a cross-cultural study has shown that patrilocal societies characteristically have only one, two, or three subsistence modes. The presence of Fraternal Interest Groups is the best predictor of feuding (Fleising and Goldenberg 1987), and, I would add, for warfare between culturally similar peoples.

Fraternal Interest Groups are the military organizations of Type A Societies. A man and his male children and their many male offspring can form a sizeable interest group. If a man's brothers and their families join with him, the Fraternal Interest Group formed is large. Cross-cultural studies from 1960 onward have shown that societies with Fraternal Interest Groups are internally violent (Thoden van Velzen and van Wetering 1960) and have frequent rape (Otterbein 1979a), feuding (K. Otterbein and C. Otterbein 1965), and warfare with culturally similar neighbors (internal war) (Otterbein 1968b).

In a study titled *Leadership, Violence, and Warfare in Small Societies*, anthropologist Stephen Younger found that the "rate of internal feuding correlated to the rate of warfare in societies with single centralized authority" (2005:107). In other words, feuding and internal war go hand-in-hand. In the same volume he attributes the violence and warfare to "male status competition" and "fraternal interest groups" (2005:125). Elsewhere, Younger argues that societies with larger populations, such as in Hawaii and Tahiti, which are maximal chiefdoms or inchoate early states (see chapter 6), have frequent wars but low levels of internal violence (2008). Leadership presumably suppresses the internal violence, an argument my wife and I made in our study of feuding (1965). Societies with Fraternal Interest Groups are said by William Divale and Marvin Harris to each have a "male supremacist complex" (1976). Mistreatment of women is part of this complex. Recognition of the complex grew directly out of Fraternal Interest Group theory. The Fraternal Interest Groups employ the basic pattern of combat described earlier—ambushes and lines. Most of the cultures mentioned so far in this book are Type A Societies, that is, Dani, Yanomamö, Higi, Andaman Islanders, and early pre-chiefdom Zulu. The Cheyenne and the Iroquois are not. They are Type B Societies, to be described next.

Type B Societies—Village Militias

Type B Societies differ from Type A Societies in a major way: they do not have Fraternal Interest Groups. Otherwise, in many ways they are the same. They are equally ancient; their people live in camps or villages and have headmen and councils of elders; and they subsist on horticulture and fishing. Unlike Type A Societies, they are usually small-game hunters and gatherers, rather than big-game hunters and herders. There are exceptions, of course, to these generalizations. The Cheyenne of the Great Plains, a Type B Society, rode horseback (like herders) and hunted bison (large game). The similar Sioux and Crow, however, were Type A Societies. The Cheyenne did not have Fraternal Interest Groups, but the Sioux and Crow did.

Type B Societies, because they do not have Fraternal Interest Groups, are relatively peaceful internally and infrequently go to war. They do have military organizations and will fight if attacked, but in comparison with Type A Societies they are not vying as aggressively for dominance. Type B Societies are characterized by cross-cutting ties. Such ties arise if men, after marriage, move to their wives' villages, a practice known as *matrilocal* or *uxorilocal residence.* Men who are unrelated to each other live in their wives' villages; polygyny will not occur unless it is marriage to two sisters—sororal polygyny.

Cross-cutting ties can also develop through the creation of different kinds of organizations within a society. Examples: teams that engage in sports, such as ball games or log racing; clubs or ceremonial organizations that put on performances; religious groups; fishermen who use the same boat; and burial societies. When two or more sets of activity groups occur in one society, the possibility exists for no two men to be together in more than one team or club. However, if the same group of unrelated men belonged to the same ball team, the same ceremonial group, the same religious organization; fished from the same boat; and belonged to the same burial society, they would form a non-Fraternal Interest Group. Cross-cutting ties are the flip side of Fraternal Interest Groups and prevent the formation of Fraternal Interest Groups. In societies with Fraternal Interest Groups, activity groups do not develop because related males probably already do many or all of these activities together—they play ball, conduct ceremonies, fish or hunt together, and so on.

The Cheyenne had numerous cross-cutting ties. A man pitched his teepee next to his father-in-law in his wife's band; his soldier society might have differed from that of other males in the band where he lived; nonsororal polygyny did not occur; and he might belong to the Council of 44, the governing body for all Cheyenne bands (Hoebel 1978:26–61). Another famous people with cross-cutting ties are the Iroquois of New York State and Ontario. They have a *clan* system (with eight matrilineal clans based on *matrilineal descent*), *matrilineages, matrilocal residence* that places related males in villages spread over hundreds of miles, and a council that draws its membership from all five separately named nations of the Iroquois (Morgan 1962[1851]).

The military organization of a Type B Society is a militia, based on village membership rather than kinship. All able-bodied men are expected to defend the village. If the headman and/or council wants retaliation for a raid, the young men who are competent fighters will be expected to participate. Loyalty to the village, not kinship, binds warriors together. Blood revenge for a homicide probably plays no role. Though *defense/revenge* is a basic or universal goal of war, village militias are likely to be pursuing *defense,* while the Fraternal Interest Groups are likely to be pursuing *revenge.*

Based on discussions with fellow **ethnographers** and reading about fieldwork experiences of ethnographers of old, I believe that it is far more pleasant and productive to conduct fieldwork in a Type B Society than in a Type A Society. Males in Type B Societies do not constantly challenge the anthropologist or other visitor. Female ethnographers would be far safer in a Type B Society. In a Type A Society they would likely be in danger—rape being a major possibility.

Early States—Elite Warriors

The term "Early State," in anthropological usage, refers to a newly formed **state** whether it arose and developed recently or existed 5,000 years ago, the approximate date for the formation of the first states in Mesopotamia (Claessen and Skalnik 1978). Another similar term, perhaps more familiar to archaeologists, is "Archaic State." "Archaic states were societies with (minimally) two class-endogamous strata (a professional ruling class and a commoner class) and a government that was both highly centralized and internally specialized" (Marcus and Feinman 1998:3–4). Such states existed in the Old World 2,000 years or more ago, and Early States developed in the New World before European contact (Marcus and Feinman 1998). An Early State is a state that is in its earliest stages. It may consist of three or four **hierarchical levels;** there will be a "capital" town or city with smaller settlements under its control (Otterbein 1977). Such a state may transition to a Mature State. From empirical research, I concluded that the point of transition is about 200 years (1986:79–80). A new, larger sample might change that figure. Early States are despotic states. They torture their own citizens as well as war captives in order to terrorize the population of the state and to terrorize their enemies. Their executions include torture in order to make death as painful as possible.

The leader of such a state is a despot or **dictator.** He comes to power by eliminating rivals—men who also want to become dictators or whom he thinks do. He is most likely a warrior, a man skilled in arms who can kill a rival in combat. To obtain the leadership role he needs help. Support comes from fellow warriors and relatives who want to become members of the ruling class. If they win, they execute their rivals; if they lose, they are executed. Mutilation may occur. Knives are used to cut off heads and to remove other body parts, such as genitals. Disemboweling also occurs. The archaeological record for some early

Archaic States provides evidence of mass executions of fellow citizens (Otterbein 2004). Ethnographies of Early States provide examples, as do accounts written about some countries in the modern world. Shaka, the ruler of the Zulu after Dingiswayo, is a famous example of an African despot, and there are African rulers in the 21st century who behave similarly. "African Despotism" is the term used for such political systems (Murdock 1959).

The supporters of the despotic leader are skilled in arms. As members of the ruling class they have the wealth, or access to wealth, that permits them to have the finest weapons, either locally made or imported. Weapons can be custom-made to a warrior's size and specifications. Shaka designed a short stabbing spear and had it made for his own use by a blacksmith. Soon the men under his command had similar weapons. When he became the king, he outfitted his entire army with the short spear. In some Early States the warriors have weapons based on individual preference, and they may choose their dress. The preferred weapon is likely to be a club, which is used in hand-to-hand combat (Otterbein 2004). Shields and helmets vary in style. Warriors are readily identifiable on the battlefield. A colorful example is Zapotec warriors; they wore helmets that resemble the head of a fierce animal, such as a puma, jaguar, or raptor; their adornments and body armor would make them winners at a Halloween party or masquerade ball (Marcus and Flannery 1996). The Zapotecs probably formed the first state in Mesoamerica (Otterbein 2004:121–130).

The *Iliad* describes, in the most vivid manner, the combat of elite warriors—their personality, their weaponry, their opponents, and their fate. A great warrior from the time of the *Iliad* gained great status/honor by killing another great warrior. Killing a lesser opponent and organizing other warriors also increased standing. Killing in close combat with a spear or sword counted more toward honor than killing with bow and arrow. Projected missiles added risk to combat, so even a great warrior could be killed by a lesser opponent. The victor in individual combat took the weapons and armor of the man whom he had killed. Sometimes disputes arose between warriors over who did the killing, and hence who was entitled to take the weapons and armor. The use of body armor and shields allowed the victor to acquire valuable commodities, which would serve as trophies and plunder. Thus, honor and plunder seem to be goals in this type of warfare (Ducrey 1985; Lendon 2005).

Iliad-style combat carried forward into the 20th century. Sniper "duels" and the aerial combat of fighter pilots have many of the same characteristics. The adversaries are known to each other, and honor comes from killing a well-known enemy, such as a deadly marksman or a multi-ace fighter pilot. (In World War II and more recently, shooting down five enemy planes gives pilots the status of ace; with 10 enemy planes one becomes a double ace.) The more kills, the more honor.

Sniper Carlos Hathcock had 93. Pilot Manfred von Richthofen, the Red Baron of Germany, had 80 kills before he was brought down by ground fire or by an allied warplane in World War I. Weapons are taken from the field of combat. Hathcock got the Vietnamese sniper's rifle; fighter pilots, like Robert Scott in the China theater in World War II, got the Samurai swords from the Japanese pilots of Zeros they downed (Scott 1943). The pilot landed and retrieved the sword, or locals recovered it for him. Both snipers and fighter pilots see whom they are killing even if the victim does not see them. Since ambush is used as in shoot-on-sight tactics, these encounters are not duels (see chapter 4). They are like the gunfights of feudists who will kill their opponent without giving him a chance. What is the point of this comparison? In my view it highlights the similarities in certain kinds of warfare regardless of time or place.

Mature States—Conscripts

Mature States arise out of Early States. They come into being approximately 200 years after the founding of the state (Otterbein 1986). The two centuries or more allow the leaders of the polity enough time to legitimize the government. The relationships between parts of the government become routinized; the constant internal strife of the Early State is gone and power is rarely needed to maintain order. For Claessen and Skalnik a Mature State develops from a Transitional Early State (1978). I have elaborated on characteristics of this type in *The Ultimate Coercive Sanction: A Cross-Cultural Study of Capital Punishment* (1986). Many states today are not Mature States; that is, they came into being in the last 200 years. The United States, born late in the 18th century, seems to be making the transition today. In contrast to an Early State, people in a Mature State are not tortured for confessions, nor are they tortured to death in a public setting designed to intimidate the masses. The leader is chosen, through designation by a council, in an election by the adult population, or from a hereditary line. There is a judicial system, with courts and judges, that strives to be fair. The polity will be large with three, four, or even five hierarchical levels, a capital city, subsidiary cities, towns, and villages. People owe allegiance to the state and its leaders. One requirement of citizenship will be service in the military.

While Mature States engage in warfare, they rarely engage in the nearly continuous warfare that engulfs Early States. Mature States, however, if attacked, will defend themselves with their long-established military and may strike back with vengeance. They may annihilate their opponent or incorporate the land and people into their developing empire as part of the polity or as a separate vassal state. Mature States have the capacity to remain at peace with their neighbors. The checks and balances within most Mature States keep them from building a military organization so large that its support is a burden on the free, nonterrorized citizenry. If the leader and his family or his supporters see possession of neighboring territory with its resources as desirable, the polity may attack. But trade, exchange or purchase of land, or a negotiation may achieve the desired ends, making war unnecessary and allowing peace to continue. Here I am defining peace as the absence of war. In the last chapter of this book I will discuss Positive Peace and the actions that can lead to achieving this goal.

The military organization will be composed of officers and conscripts. In many Mature States, men, rarely women, have an obligation to serve in the military and can be called up or drafted as needed. Since the polity is likely to be large, the military organization may be large. The recruits will be organized into similar units and led by officers, who are likely to come from the upper class. If the polity is a monarchy, officers may be sons or relatives of the king. Indeed, the monarch himself may be the supreme commander. The officers may have had training in the use of weapons, both personal arms and more specialized equipment such as chariots and catapults (artillery). It is important that they have learned to lead and command since they need to direct the units on the march and battlefield. Learning maneuvers is essential since the armies march in columns and form lines on the battlefield. Similar units in line are the second type of line described in chapter 2. If there are specialized forces, such as cavalry and artillery, their deployment requires the coordination of different units on the field of battle, a difficult task for the commander and his senior officers ("combined arms team").

It is expensive for a Mature State to have an army. A major characteristic of a drafted or conscript army is that the soldiers will be dressed alike. And being dressed uniformly probably means that their clothing, armor, and weapons have been provided by the government. The provision of clothing and weapons to conscripts means that the government has to have the means to manufacture the equipment, store it, and supply it to the soldiers. Officers may be dressed differently and carry different weapons so that the conscripts can identify their officers. Officers can also be identified by the enemy and become a target for marksmen with bows and arrows or rifles. The capture of royal officers, even the king himself, may be a major goal of combat, because these high-level combatants can be either publicly executed or ransomed for great wealth.

Being dressed alike means that it is possible to identify a conscript army in paintings, drawings, or sculptures of battles. Over four thousand years ago in Mesopotamia, artwork showed similarly dressed soldiers on the Standard of Ur, which is an artifact that dates c. 2500 BCE. It shows battle scenes of the Sumerian army, in which warriors wear capes and helmets, carry spears, and follow four-wheeled carts

Political Systems	
Uncentralized Political System	**Centralized Political System**

Type A Societies	**Type B Societies**	**Early States** →	**Mature States**
• Villages or settlements of dispersed homesteads run by headman/ council	• Villages or camps run by headman/ council	• Despotic leader	• Monarch or representative leader
• Patrilocal	• Matrilocal, multilocal	• 3–4 hierarchi-cal levels	• 3–5 hierarchi-cal levels
• Polygynous	• Monogamous	• War frequent	• War is often unnecessary due to trade, negotiations, and purchase of land
• Ties based on kinship	• Cross-cutting ties; ties based on loyalty to village	• Internal strife	
• Internal violence (feuding)	• Internal harmony		• Fair judiciary

Military Organizations

	Nonprofessional			Professional	
Type A Societies	**Type B Societies**		**Early States**	**Mature States**	
Fraternal Interest Groups	Local Militias		Elite Warriors; ruling class over commoners	Conscripts; officers come from upper class	

— Raids —	— Campaigns —
Informal Tactics	Formal Tactics

Ambushes & Line Battles	**Line Battles & Sieges** (Integrated tactical systems)
• Spared women and children	• Didn't spare women and children
• Women became mates	• Took captives as slaves
• Some took captives and sold them to centralized political systems	• Sold captives into slavery

Goals of War

Type A Societies	**Type B Societies**	**Early States**	**Mature States**
Plunder Prestige		Plunder	Defense
Revenge and Defense		Political Control	

Figure 3.1 Characteristics of Uncentralized and Centralized Political Systems and Their Military Organizations

pulled by equids. Enemy lie dying on the battlefield. In other scenes, the captured enemy are paraded before the monarch. Another item from Ur, only slightly more recent, is the Steele of Vultures. Again, similarly equipped soldiers formed into phalanxes march over the bodies of fallen enemy. The best photographs of this artwork seem to be in Yadin (1963:132–135); see also Otterbein (2004:151–156) for detailed descriptions of the battles depicted.

To summarize, the four polities and their military organizations can be displayed in figure 3.2, which emphasizes the degree of centralization of the types and the extent of warfare inherent in each:

	Warfare Common	**Warfare Uncommon**
Centralized Political Systems	Early/Despotic State (Elite Warriors)	Mature State (Conscripts)
Uncentralized Political Systems	Type A Societies (Fraternal Interest Groups)	Type B Societies (Village Militia)

Figure 3.2 Types of Sociopolitical Organization Related to War and Peace

WHY POLITIES AND THEIR MILITARY ORGANIZATIONS GO TO WAR

The question of why people go to war has been answered in two different ways. One answer focuses on the intentions and goals of the political leaders and the military organization itself. The other answer focuses on the underlying conditions, whether they are generated by subsistence needs, changes to the physical environment, the structure of the sociopolitical order, or cultural beliefs. Although people are often unaware of the underlying conditions, they know the goals of their immediate actions, no matter how irrational other members of government or military think they might be. The study of polities and war is difficult because leaders may conceal their true goals by enunciating acceptable goals. Anthropologist F. G. Bailey has pointed out what he calls "pragmatic rules," which are intended to accomplish political goals, and "normative rules," which are intended to deceive an audience (1969:1–7).

Uncentralized Political Systems

Whether Fraternal Interest Groups or Village Militias, the reasons members of a raiding party go to war will be similar. Thus, there would be similar reasons for both Type A and Type B Societies. The raiders are secretly making plans at night to attack a nearby enemy village. Their individual motives and plans are rarely spoken but might include the following:

1. Headman—"If I can kill 'X' (headman of the other village), then I'll be the most important headman in the region."

2. Young Brave—"If I can kill 'Y,' I'll be able to seize his cute wife and bring her back to live with me."

3. Adult Male—"That bastard 'Z' killed my brother last year. If I can kill him, I'll have revenge and reestablish family honor."

4. Farmer—"If we can drive them from their village and land, I'll be able to expand my fields."

5. Old Married Man—"If I don't go I'll be laughed at. If I hold back, maybe I'll not be injured."

6. Warrior "On the Make"—"If I can take another head, I'll become known as the fiercest warrior in our village."

Centralized Political Systems

Whether Early States or Mature States, the reasons members of a war council have for sending their military organization on a campaign will be similar. The political and military leaders are at the governmental headquarters in the capital city. Deliberations are secret. Once a decision is made the general in charge of the army will be notified.

The following are reasons members of a war council order the military organization to attack another polity:

1. King, President, or Prime Minister—"The enemy has been building up their military and may attack us. We need to attack first." (This is a preemptive attack. He or she may also have secret reasons, such as wishing to gain more power or suppress dissent.)

2. Vice President or Second in Command—"This will be a disaster, but if I let this fool go ahead, maybe the legislators will replace him, and I'll become president."

3. Sole Woman on Council—"If I don't agree to the attack, I'll be called a weak-willed woman."

4. Military Chief-of-Staff—"I favor this war because the military will grow in size and become more important. Besides, the soldiers need practice."

5. Representative of Businesses (Secretary of Commerce)—"My former company will get a lot of new business from government contracts."

6. Foreign Minister—"Let us first use coercive diplomacy and try to get them to disarm, and if they do, we will not need to attack." (His secret motive may be to gain influence with the political leader.)

CAUSES AND CONSEQUENCES OF WAR

A full analysis of a warfare system must take account of four components: material or underlying causes, efficient or proximate causes, war itself, and the many consequences of war. Nearly all analyses of warfare by anthropologists take these components into account. Indeed, I have argued that over the past 50 years there has been a convergence in anthropological theories or models, such that today nearly all match the following paradigm (1994a:177).

Material Causes ⟶ Efficient Causes ⟶ War ⟶ Consequences

The terms "material causes" and "efficient causes" derive from Aristotle (1984:149–155). They seem to be more commonly known to anthropologists as "underlying" and "proximate" causes. I use the expression "goals of war" for the latter (1970): the explicit reasons given in ethnographies by informants are the proximate causes or goals of war. Underlying causes are derived by the analyst. It is an inductive approach. For example, if people appear hungry, food shortage is inferred; if population density is high, crowding and its consequences are inferred; if hurricanes or storms displace people, intergroup contact, which can be violent, can be inferred; if the polity has Fraternal Interest Groups, internal and external violence are inferred; if the polity is an Early State, its internal operations are inferred to be generating violence, torture, and executions; if the people believe in malevolent gods, then coercive treatment of children, of subordinates, and of captured enemies will be inferred. The list could go on. Each of these reasons could lead to war but they need to be experienced in such a way that polity leaders and warriors will formulate goals that could correct the underlying stresses through successful combat. Andrew Vayda (1976) called such stresses "perturbations," a word I needed to look up in the dictionary. Perturbations disrupt the functioning of the system.

I look for underlying reasons in four domains: physical environment (climate/weather), resources (mineral, plant, animal), social structure (social organization, political organization), and culture (beliefs, rules and laws, and worldview). Among those who focus on underlying causes, four groups stand out: (1) those who examine the physical environment, like archaeologist Brian Fagan whose book title gives his position, *Floods, Famines and Emperors: El Niño and the Fate of Civilizations* (1999); (2) those who believe that competing over scarce resources (CSR) is *the* major cause of war—these are materialists; (3) structuralists, like myself, who believe that the social and political organization of a polity is the prime mover; and (4) those who stress patterns or themes inherent in the culture as causative and are sometimes lampooned for allegedly saying that "warlike people go to war." The last reason is Ruth Benedict's argument as to why the Kwakiutl of the Northwest Coast engaged in aggressive warfare (1934).

My list of goals that politics hope to achieve through war contains four groupings. (1) defense and revenge, (2) plunder and land, (3) trophies and honor, and (4) subjugation and tribute (1970). The greater the political complexity of the polity, the more reasons will be given for going to war. Thus, a small-scale single band or village polity may go to war only for defense and revenge, while a state may go to war for all four goals.

The consequences of war are likely to involve territorial expansion or contraction. Even if the goals of war for the attacking, successful polity do not include land, the defeated may flee and leave land available for occupation by the victors. This seems to have occurred for the Dani and other highland New Guinea groups (Ember 1982). In *The Evolution of War* I found that the best measure of military success was territorial expansion regardless of what the stated goals had been (1970). A consequence of war that was discussed in chapter 2 is casualties to both the warriors/soldiers and the civilians. Villages are destroyed and burned, as well as other resources such as food stored in granaries. If soldiers cannot take grain with them, they typically set fire to the granaries. The American army during the Revolutionary War defeated the Seneca Iroquois in present-day western New York by burning the summer's harvest in a fall campaign led by General Sullivan (Emerson 2004). Other destruction may occur, such as tearing down dams, chopping down fruit trees, and polluting wells and waterways. Not all consequences, however, are bad. Some anthropologists have studied buffer zones that arise between warring populations. Animals, including rare species, may flourish in these zones (Hickerson 1965). A modern example is the DMZ (demilitarized zone) between North and South Korea where rare birds and animals have now been reported. But on balance, war is destructive to both winners and losers and should be prevented if it can be (see chapter 8).

The final two sections of this chapter have presented different approaches to the study of the causes of war. The four-stage paradigm is intended to tie the two approaches together. One focuses on underlying causes, the other on the goals that those who have the ability or right to go to war seek to achieve. Those who focus on underlying causes see those who focus on goals as naïve, because people do not know what motivates them. Those who focus on goals see those who look for underlying reasons as naïve because they do not realize that they are setting forth ex post facto explanations, as economists do when they explain the next day why the stock market went up or down, but the previous day could not accurately predict the direction of the market. Science of uncertainty expert Nassim Nicholas Taleb has referred to this as "back-fit logic" (2005:xvi) or "retrospective predictability" (2007:xviii). I believe that many anthropologists today are combining both approaches.

Chapter Four

War and Its Cousins

Numerous anthropologists, other social scientists, and other interested persons seem to be satisfied with considering warfare as violence; they do not require a precise definition. But I do. To lump warfare together with all other types of killing forces researchers to seek psychological explanations for war. The classic explanation for warfare as violence comes from Sigmund Freud in "Why War?". Freud uses the term violence and follows with the idea that humans have in them an active "instinct for hatred and destruction" (1959[1932]). This perspective continues today. A textbook titled *Violence and Culture* by Jack David Eller calls war "polity-versus-polity violence" (2006:214–216). Eller makes no distinction between war and violence; "war" is in the index only in entries for World War I, World War II, and World War III.

I will define war and other types of homicide so that each type can be distinguished and identified, just as a bird watcher can distinguish a bluebird from a tree swallow, a barn swallow, or a purple martin by knowing unique features of each. I believe each type of homicide is distinct, each has a separate origin, and there are separate ways to prevent each. To consider an "instinct" to be the cause of all killing, and then group it in one category labeled "violence," is a mistake. Before examining this approach to homicide, we will explore "warfare" in more depth.

TYPES OF WARFARE

Warfare is armed combat between political communities. Armed combat, as noted in chapter 2, is fighting with weapons. But the term **political community** has yet to be used, because I have used the better-known term "polity." I use these terms interchangeably. Comparative anthropologist Raoul Naroll provides a definition of political community: "A group of people whose membership is defined in terms of occupancy of a common territory and who have an official with the special function of announcing group decisions—a function exercised at least once a year." It is his definition for a "territorial team" (1964:286). I use the definition because it is applicable to uncentralized political systems. A further definitional requirement of a political community is that it is a *maximal territorial unit;* that is, it is not included within a larger unit. Eller states: "A *polity* is a sovereign political entity, usually but not always a state" (2006:213). Yes, states are polities, but uncentralized political systems are also polities, from bands to villages.

Each political community has a leader or leaders. In an uncentralized political system they are usually called headmen. The *headman* of a band or single-settlement political community will probably be an *informal* leader with a *limited* degree of power. If multiple settlements form the polity, the political community is considered to have two hierarchical levels. More complex polities are composed of three or four levels. The *political leaders* of two-, three-, and four-level polities are referred to as *chiefs, kings,* or *dictators.* The following figure 4.1 summarizes each type (Otterbein 1977:130).

Types of Political Leaders	Degree of Formality	Degree of Power
King	Formal	Absolute
Dictator	Informal	Absolute
Chief	Formal	Limited
Headman	Informal	Limited

Figure 4.1   Political Leaders and Their Degree of Formality and Power

There is no leaderless political community. Even the smallest band or politically independent settlement composed of only a small family or *household* will have a leader. The father or head of household will be the leader, and he can be called a headman. Some may prefer to call the leader a "headperson," although it is highly unlikely that a woman will be the leader (for an exception see Endicott and Endicott 2008).

More definitions are needed. The key concept of anthropology is *culture,* the way of life of a particular group of people. And *a culture* is a particular group of people who share the same way of life. Two criteria are used to distinguish one group of people from another: language distinctness and geographic separation. If different peoples speak different languages, they are different cultures.

The use of our two key concepts, political community and a culture, leads to a distinction between two types of warfare, internal war and external war (Otterbein 1966). This pair of terms may have become standard usage in anthropology; they have proven useful to warfare researchers. *Internal war* means internal to a culture. For such war to occur there must be two or more political communities whose members speak the same language. When they are at war, it is war within the culture, or internal to the culture. (Political scientists use the term internal war to refer to war within a polity. For them, insurgency or revolutionary wars are called internal war.) *External war* takes place between political communities that are culturally different; different languages are spoken (Otterbein 1968b).

The distinction is important: if war occurs between two polities in the same culture, the weapons, tactics, and other elements involved will be the same; if war occurs between two polities that are culturally different, the weapons, tactics, and so forth are likely to be different, and possibly greatly different. Imagine how lopsided the battles may be if a Type A or a Type B polity is at war with an Early or Mature State. Not only may the weapons systems and military organizations be different, the goals of war may be different, and what war means to each culture may be incomprehensible to the other side. In internal war it is difficult to know from reading an ethnography who was the attacker. In external war it may be easy to identify the attacker since the size and military abilities of each polity may differ greatly. Thus, I distinguish external war—attacking (offensive war) from external war—being attacked (defensive war). For any war that can be called external war, one side will be the attacker, the other side the defender. The qualitative difference between attacking and defending has made this pair of terms useful.

The various distinctions I have made between types of warfare were developed for my use and for fellow anthropologists. Historians, political scientists, and the media, on the other hand, use terms applicable to modern warfare. Nuclear warfare developed suddenly at the end of World War II when a single bomb destroyed a Japanese city. The land and sea warfare that had been engaged in by armies and navies since the time of the Ancient Greeks came to be known as conventional warfare. Mountains of books have been written to assist commanders and political leaders in the "art of war." For example, see Gèrard Chaliand, *The Art of War in World History: From Antiquity to the Nuclear Age* (1994). Means of preventing either a nuclear or conventional attack have been discussed under the concept of deterrence, which is described in chapter 8.

A third type of modern warfare has taken a central role in military thinking—guerrilla warfare. In 1961, when I developed a serious interest in the study of war, I posed the question: is primitive warfare guerrilla warfare? Because scale and tactics were similar, I thought the answer might be "Yes." Indeed, before entering academia, where I would be able to influence young minds, I took a position as a researcher at a "think tank" in Washington, D.C. There, I learned that guerrilla warfare is *not* primitive warfare, but rather it is the lethal combat used by insurgents and terrorists. The goal of a guerrilla army, such as the one led by Fidel Castro, is to overthrow the government of the combatants' country, nation, or polity. Castro's lieutenant, Che Guevara, had written a manual on how to do it (1969). The U.S. Army wrote manuals on how to prevent guerrilla warfare. In 1962, Field Manual 31-16, titled *Counterguerrilla Operations* (Department of the Army), was issued to troops, and I worked on the topic "civic action," popularly referred to as "winning the hearts and minds of the people." The media today write and speak as if counter-insurgency is a recent development spearheaded by generals. This is not so. The new U.S. Army Field Manual, No. 3-24, is titled *Counterinsurgency,* with Forewords by Gen. David H. Petraeus, Lt. Gen. James F. Amos, and Lt. Col. John A. Nagl (Department of the Army 2007). Social scientists Sarah Sewall and Montgomery McFate assisted in the writing.

The Human Terrain Teams (a program begun in 2003 that embeds social scientists with combat brigades) are a descendent of decades of thinking about how to prevent insurgents from winning. Trained human intelligence (HUMIT) collectors are soldiers who hold military occupational specialties (Department of the Army 2007). The use of civilians, including anthropologists, however, has become highly controversial. A review of the complicated relationship between the Department of Defense and anthropologists from World War I to the present is detailed by anthropologist Montgomery McFate (2005). Between the 1960s and 2005, the relationship went from positive, with the U.S. Army wanting the assistance of anthropologists, to a parting of company, to the generals today embracing anthropologists.

From this insurgency/guerilla warfare/counterinsurgency setting has emerged another type of warfare referred to as "terrorist warfare," which entails a world of cultures in conflict. "Wage jihad and kill Americans" is the rallying cry of terrorists who want to destroy all Western governments and nations and their allies. Ralph Peters describes "five social pools" from which most of these killers emerge: the underclass, young males deprived of education, entrepreneurs of conflict, patriots, and failed military men. The entrepreneurs and the patriots, he believes, pose the greatest danger to our social order; the others are swept along by the tide (1999). Western forces are developing high-tech gear designed to dismantle opposing forces through "precision violence." Such pinpointing of terrorists contrasts with what I call "diffuse violence" or free-ranging violence. What I observe daily is the precision violence of NATO forces and the diffuse violence of the terrorists. If the West, however, reverts to the genocidal practices of World War II and employs diffuse violence against the populations from which the terrorists arise, tens of millions will die.

Negotiations take place among the most simple societies, as they are among the more complex societies that are described above. It takes only one political leader from each side, or the leader's representative, to negotiate. Two polities that are about to engage in war may enter into negotiations when one polity is about to begin siege operations against a city (chapter 2). Here, coercive diplomacy may be employed in an attempt to obtain concessions or surrender. When a battle or war has ended with one side the victor, the victor will be able to enforce its demands. In contrast, political communities may negotiate in an attempt to create an alliance; these negotiations will differ greatly from those that take place between polities on the verge of warfare. A common language may facilitate the formation of an alliance. Generally, alliances are about war or peace—an agreement to attack another polity or to come to one another's aid if one polity is attacked. Both the Dani and the Yanomamö polities form alliances for the purpose of attacking another polity. This ability of leaders and their polities to form alliances can be a major generator of warfare. Alliance warfare, as described previously for the Dani and Yanomamö, can result in huge casualties for the loser.

SEVEN COUSINS OF WARFARE

The purpose of this section is to identify and define types of armed combat and killing that have often been lumped together with warfare. Most people in most cultures view them as different, both in form and function. While the seven types described below can be classified together as forms of killing, their origins and outcomes greatly vary. They cannot be explained by a single psychological theory, or even by a number of psychological theories.

Self-Redress

Self-redress occurs in situations where a person, usually a man, feels wronged. To right the wrong he takes action against the person whom he blames. This action has been called self-help, coercion, or folk justice. Self-redress has been defined as one "disputant takes unilateral action in an attempt to prevail in a dispute or to punish another" (Fry 2006:23). Self-redress takes place generally in an uncentralized political system without Fraternal Interest Groups and feuding or without a formal legal authority who can intercede on the victim's behalf. The action often consists of the disputant taking back what he felt to be his own or of injuring and even killing the party who wronged him. The community sees the action as justified. This is why it is called "folk justice" by William Montell (1986). Self-redress seems to be restricted to Type B Societies, since in Type A Societies Fraternal Interest Groups are the contending parties.

Crimes can be grouped under three major categories: theft, personal crime, and communal crime. The first two pertain to the individual and can lead to an attempt on the part of the victim to recover what was stolen or to punish the attacker. In political communities with kinship groups, such as Fraternal Interest Groups, the victim's kinship group may assist him and thus set off a feud (see "Feuding," below; Otterbein 1996b). Personal crimes can include assault, attempted murder, a sexual offense or rape, sorcery intended to make another ill, and false accusations. Personal crimes against an immediate family member are also grounds for self-redress. Communal crimes are dealt with by the community at large, sometimes by the execution of the culprit (see "Capital Punishment," below; Otterbein 1996a).

Self-defense is a form of self-help. The intended victim fights back by attempting to frighten, injure, or kill the assailant. Personal weapons, such as a knife, club, or firearm, are carried by individuals in many cultures. The Higi carry the "f" shaped slashing weapon. Martial arts expertise is a weapon that can be concealed and go everywhere with those who possess it. Many peoples have developed martial arts that are particular to their cultures (Green 2001). The handgun is another type of weapon that is easy to conceal and is portable. Even in modern states considered to be democracies some citizens will be given the right to carry concealed handguns. If law enforcement determines that an individual was acting in self-defense, any homicide committed with the pistol or revolver will not lead to charges and conviction (Kennett and Anderson 1975).

A variant on individual self-redress has been reported in at least one ethnography. The Comanche Indians of the American Southwest were studied by E. Adamson Hoebel in the 1930s. In previous decades the Comanche were a hunting-gathering band famed for its bellicosity. Hoebel studied what he called the law-ways of the Comanche. Their legal system was rudimentary. If one were wronged by a strong adversary he would obtain the help of a Champion to get justice. The Champion was more like a "hired gun" than a lawyer. He would not take one's enemy to court—there were none—but go after him with weapon in hand. Like a lawyer, the Champion charged his client a fee for his services (Hoebel 1940, 1954).

Homicide and Political Assassination

Homicide and political assassination occur within a political community. Homicide is the intentional or unintentional killing of another person. The killer and his victim are both members of the same political community or else the political leader controls the region in which the killing has occurred. Political leaders do not approve or regard homicides as appropriate. In most cultures, homicides, particularly intentional killings, are crimes. The killer may be subject to severe punishment, such as banishment, long-term incarceration (if jails or prisons are present), or capital punishment (Otterbein 1986; 1996a).

Political assassination is the killing of the leader of the polity, or a person appointed by the leader, by a member or members of a political community. If the assassination is performed by someone who belongs to another political community, the act can be regarded as warfare. Political assassination is always a criminal offense unless the political leader himself or herself arranges to have a subordinate leader killed (Otterbein 1986). Others may view this as a crime, but generally they can do nothing about it except overthrow the government. If the hired killer is caught, the political leader can later pardon him or her. Political assassination, thus, may lead to war or revolution.

Feuding

Many scholars have despaired of separating feuding from warfare, and as a consequence have lumped feuding and warfare together as internal war (Ember and Ember 1971; Ross 1993). Some scholars have classified what I believe to be cases of warfare as feuding (Schneider 1950; Fry 2006; and Sponsel 1998). I maintain that a single criterion can be used to distinguish one from the other. Warfare is armed combat *between* political communities, while feuding is *within* political communities. In order to apply the distinction one must be able to identify political communities. I have found this easy to do for nearly all cultures and their polities except for small scattered peoples where the identification of the political leader or headman is difficult. I have found Naroll's (1964) definition can be applied in nearly every situation. If someone announces decisions, he is the leader.

When a *local group* is both a kinship group and a political community, the armed combat—ambushes and line battles—is considered feuding if the focus is on the social organization, and it is considered warfare if the focus is on the political community. My view is that the political community takes precedent. The constitution of the fighting units does not prevail. If the local groups are culturally the same and the armed combat is viewed as being between political communities, it is internal war. For decades social scientists have disagreed about whether the Australian Aborigines engaged in feuding or warfare. An early scholar, Schneider (1950), focused on kinship and concluded that it was feuding. Recently Fry (2006) has done the same. He has further considered feuding as a legal process. Thus, the best known Australian peoples, the Aranda, Murngin, and Tiwi, are classified as having feuds, not warfare. I have classified the Tiwi as having warfare (Otterbein 2004), and I would also classify the Aranda and Murngin as people who go to war, because I focus on their political communities.

Feuding is a series of revenge-based killings that contribute to the disruption of the social order. There are five essential elements to a feud. The first is the all-important distinction made above. (1) The feud takes place within the political community, whether it be one of the four types of political systems identified in chapter 3 or a modern nation belonging to the United Nations. (2) Kinship groups are involved. These are often Fraternal Interest Groups, but other kinship structures may be involved. Northern Luzon in the Philippines is famous in anthropological circles for its feuds, yet Fraternal Interest Groups are not present. The Ifugao and Kalinga have *kindreds* that have developed an elaborate system of feuding with many rules and an elaborate schedule of *compensation payments* (Barton 1919, 1949). The Kalinga also engage in warfare. They go on head-hunting expeditions. The heads taken increase the fertility of land, animals, and women (Service 1963:271–290 provides a brief description). (3) Homicides take place. A series of counterattacks and only woundings can also constitute a feud sequence. (4) Three or more alternating killings or acts of violence occur. (5) The killings occur as revenge for injustice. The terms *duty*, *honor*, *righteous*, and *legitimate* appear in discussions of the motivation for the homicides (Otterbein 2000b). Note how similar feuding is to individual self-redress. With feuds, kinship groups attempt to rectify the offense; with self-redress, the individual is on his own unless he can employ a Champion or find a "hired gun," a favorite theme of American and Japanese movies.

Napoleon Chagnon's analysis of Yanomamö warfare has made several shifts over the decades. When Chagnon first conducted fieldwork with the Yanomamö in the 1960s he used theory from international relations research and concluded that the Yanomamö fought to maintain political sovereignty. He definitely viewed Yanomamö raids and the treacherous feasts as *warfare*. To combat the attacks of materialists who argued that Yanomamö fighting was due to protein scarcity, in the 1970s Chagnon focused on a new reason for war—fighting over women; if a kin group could increase its population of women, it could produce more male offspring and become larger and stronger. By the late 1980s he focused on the kin group, not the village, as the unit that fights. Revenge became the new and third reason for war. Or is it still war? With revenge being featured, Leslie Sponsel concluded that Yanomamö warfare was feuding (1998), a conclusion I do not agree with. By the 1990s new field research in highland villages led Chagnon to conclude that the Yanomamö fought to occupy the resource-rich lowlands (Otterbein 1994a:175–176; 1994b:219). Although Chagnon could be charged with being a "theoretical chameleon," I commend him for being intellectually flexible. I believe the four reasons identified for war can be fitted to the paradigm presented near the end of chapter 3. Maintaining political sovereignty is an underlying cause, and I would add presence of Fraternal Interest Groups and the alliance system to a list of underlying causes. Fighting over women and for revenge are proximate causes. Yanomamö movement into the lowlands is a consequence of successful warfare.

While I have no difficulty in classifying Yanomamö armed combat as warfare, our next example is challenging. Ethnographer Rolf Kuschel conducted extensive field research on Bellona Island, a Polynesian outlier, which had almost no prior contact with the outside world. In terms of degree of contact with Western civilization, the Bellona are more like the Dani than the Yanomamö or Higi. Kuschel describes Bellona armed combat as "blood feuds." Some of these feuds went back three centuries (1988:170). He believed that the population of the island seldom exceeded 500, but the island could have fed about 1,000 persons. Neither Kuschel nor the Bellonese felt that the killing resulted from overpopulation. Bellona clans were politically autonomous Fraternal Interest Groups, which leads me to classify their "feuds" as warfare (Kuschel 1988:20, 61). The clans did not pay compensation to end the "feuds." Further support for my interpretation arises from Kuschel's contention that the intent of a raid was not to even the score but to reduce manpower of the other side. Once a pattern of mutual raiding was set in motion, it was likely to continue until one group was nearly annihilated. Groups small in size hid in the bush so that raiding parties could not find them. Bellonese warfare seemed to be effective in reducing population numbers.

Most accounts of feuding assume a tit-for-tat feud sequence, where killings and counterkillings alternate. We saw in the Dani case study that while each battle was an attempt to even the score, one side eventually pulled ahead. Although this was war, not feud, the same process applies to both. In my study of five Kentucky feuds, I found that the same lopsidedness sometimes developed. In the Turner–Howard feud, 13 Turners died while no Howards were killed (2000b).

Dueling

Dueling has a close relationship to individual self-redress. A duel may arise when a man who feels wronged challenges his enemy to a duel and the person accepts. The political community or the segment thereof in which they reside, however, must have the practice or institution of dueling, Most nonliterate cultures do not have dueling; there are few exceptions. Duelists fight with matched weapons, which are lethal. They agree on conditions such as time, place, weapons, and who should be present. They are from the same social class. Motives range from preserving honor to revenge to the killing of a rival, with honor most frequently mentioned (Otterbein 2001).

Dueling may in some cases have a close relationship to war. Although duels usually take place within the same political community, with the duelists being from the same culture and social class, duels do take place between duelists from different military organizations. The combatants—they are elite warriors—stand in front of their respective military organizations. The Zulu had what I have called dueling battles. Chosen warriors, Shaka being one, would advance to within 50 yards of each other and shout insults. The warriors carried five-foot-tall oval shields and two or three light javelins. The combat opened with the warriors hurling their spears—a duel. Shaka charged with his short broad spear and stabbed his opponent in the left side. This seems to no longer be a duel. The combat between David and Goliath preceded a battle, but since David used a sling (a projectile weapon) and Goliath a sword (a shock weapon), I would not classify the combat as a duel. Plains Indians fought duels. The artist George Catlin described a duel between a Mandan leader whom he painted and a Cheyenne war chief. They used the same weapons throughout the fight. When down to only knives, the Mandan killed and scalped the Cheyenne. A duel or individual combat preceding a battle is generally the opening phase of an all-out battle (Otterbein 2001).

Dueling and feuding, however, are culturally incompatible with each other. Feuding works against the development of dueling within polities. In feuding societies honor focuses on the kinship group, in dueling societies, it focuses on the individual. In a feuding culture, no one would dare intentionally kill or injure another in a duel. If a duel occurred in a area where feuding was an accepted practice, the resulting injury or death would start a feud between the kinship groups of the participants. In other words, dueling neither develops in nor is accepted by feuding societies. Where there are feuds, there are no duels (Otterbein 2001). In the American lowland South, dueling flourished. Feuding flourished in the hill country (Otterbein 2006a). Two famous early leaders fought duels, one surviving to become the seventh president, the other dying before he could become president. (If you have trouble remembering who they are, look at the folding money in your wallet.) A major center for dueling was the Low Country of South Carolina, a land of slave plantations (Long 2008).

Capital Punishment

Capital punishment is defined as the "appropriate killing of a person who has committed a crime within a political community" (Otterbein 1986:9). There are three aspects to the definition: (1) "The killing occurs within a political community or within a small unit of the political community. . . . 'Within' means that the execution takes place within a geographic area controlled by a political leader; although the person who is to be executed is likely to be a member of the political community this is not necessarily the case—the person may belong to another political community" (Otterbein 1986:9). Two cases from different continents will be presented shortly that involve executions that may be outside the geographic area controlled. (2) The killing is considered appropriate or has the approval of the political leader, or the leader of a smaller unit. (3) "There is a reason for the killing. Someone has performed an act (or a series of acts) that is considered to be a crime by the political leader or members of the political community" (Otterbein 1986:10).

Sometimes the political leader, for fear of angering the kin of the criminal, will employ executioners from another political community to kill the offender either within the political community, in a no-man's-land, or in the territory of the other political community. This can appear to be warfare. A raiding party ambushes and kills a member of another culture. But it is not war. Motives are entirely different, and the killing does not lead to hostilities between the polities and their leaders. The criminal's kin may want to take revenge, but they do not know the identity of the ambushers or their motives.

An early example comes from the Aranda (Arunta) of Central Australia, perhaps the most famous of the aboriginal cultures of that continent. They have one of the most complex kinship systems ever devised by Homo sapiens. It is a challenge to learn, but here it is not necessary to master it. (See Murdock [1934:20–47] and Service [1963:3–26] for summaries.) The classic ethnography by Spencer and Gillen is *The Arunta: A Study of a Stone Age People.* An individual had to choose a marriage partner from a particular kinship group (Spencer and Gillen 1927).

> To have sexual relations with a person not of the prescribed group was regarded as incest and was punished by death. The elders of the band decided in secret upon the execution and asked a neighboring band to carry out the sentence of death. (Otterbein 1986:50)

Another example from a horticultural society comes from an area in West Africa, to the east of the Higi, in the county of Cameroon. Each Meta village was governed by a council presided over by a chief. One responsibility of the council included the settlement of interlineage disputes (Dillon 1980b:661). Secrecy characterized the several steps in the decision-making process. A complaint would be brought to the chief of the offender's village, who in turn would seek the "unanimous assent of offender's lineage" (Dillon 1980a:443) to the execution of the offender, often a habitual witch or thief. Once the chief's consent was obtained, the execution could proceed. Usually chosen to do the killing would be a member of a **lineage** recently settled in the area, who would not have kin ties to the offender. A new lineage came from a neighboring political community and would have wanted to ingratiate itself to its new chief. The executioner "dressed in a hood of raffia cloth and carrying a . . . soot-blackened club" would ambush the victim and kill him. The hood and blackened club were to keep the executioner's identity secret in case there were witnesses (Dillon 1980a:445).

Human Sacrifice

Human sacrifice occurs within a political community and is deemed appropriate by political leaders. The persons who are offered to the deities have not necessarily committed crimes. However, if the sacrificial victim is either a criminal or a war captive, the killing is also capital punishment (Otterbein 1986). In a small cross-cultural study, I found that human sacrifice occurs most frequently in polities with two or more hierarchical levels, but in Early States rather than in Mature States. Among uncentralized political communities, those with councils of elders are likely to have human sacrifice, while uncentralized political communities that are bands or are without councils will not have human sacrifice (see Otterbein 2004:197–198 for tabulated data).

In a similar sample Dean Shields found a relationship between human sacrifice and the presence of slavery and full-time craft specialists (1980). Both sets of findings show that societies that are intermediate in size sacrifice both societal members and captured enemies. If criminals and war captives are sacrificed to the spirit world, the deaths can be classified as both capital punishment and human sacrifice (Otterbein 1986:12). Nevertheless, the motives for each differ; in one case the motive is to rid the political community of a wrongdoer; in the other it is to approach the spirit world. We need not linger any longer discussing capital punishment and human sacrifice, but any study of the warfare system of a set of political communities should examine the treatment of war captives (see chapter 2) and ascertain whether their execution is considered a punishment or a sacrifice to the spirit world.

Genocide

Although the origin of genocide goes back into prehistory, as does war itself and several of its cousins, the term has a precise origin point. Polish lawyer Raphael Lemkin in 1944 derived the term from the Greek *genos* (race) and Latin *caedere* (to kill). He used "genocide" to describe the Holocaust. The United Nations further defined genocide in 1946 as "a denial of the right of existence of entire human groups" (UN 1946:188). By 1948 the United Nations became concerned with the Prevention and Punishment of the Crime of Genocide. The targeted group was expanded to also cover national, ethnic, and religious groups. Since then, social and political sections of a polity have been added to the list. The definition came to include the notion that it was a despot and his government that did the killing of people within the polity they controlled. I would add that a conquering army can also intentionally kill residents of the polity attacked. Genocide occurs if a raiding party tries to annihilate an enemy village, whether its members are of the same or a different culture (Corry 1996).

Some anthropologists have extended the term to include the destruction of native populations that occurred when Eastern and Western powers expanded territorially. In their rush to become empires, they killed and relocated native peoples to the poorest land available. I have called these groups Dependent Native Peoples (1977:243). Others have called them "victims of progress" (Bodley 1982). American Indians, such as the Cherokee in the southeast and the Sioux and Cheyenne in the northern plains, who were relocated by the U.S. government to reservations, are frequently used as examples. Today, globalization is seen by some as keeping undeveloped nations in a state of poverty (Wolf 1982). A term for this phenomenon is "structural violence." I see as the key element in the definition of genocide the intentional destruction of individuals who are members of a certain category (e.g., "Jew," "Indian," "Aborigine," or "Tibetan") by a government that has either taken control of their country (e.g., Nazis in Germany) or has recently conquered the territory (e.g., the Chinese in Tibet). Genocide can be the result of war; it can also result from one ethnic group or segment of a polity attempting to take exclusive control of the government and the state and, in the process, killing as many members as they can of the ethnic group that had ruled the country.

LEVELS OF ARMED COMBAT

Sometimes some of the seven forms of killing in a particular culture are related to warfare and to each other. To demonstrate this, several ethnographers have presented the forms of killing in the polity they researched in a sequential order, like that presented above. When there are connections between some of the types, the connections are described. I have named this procedure as identifying "levels of armed combat." Three examples follow:

I believe the first such study is presented in an article by W. Lloyd Warner from 1931, which was incorporated into *A Black Civilization* (1964:144–179[1937]). The Australian Aborigines, the Murngin, whom Warner studied resided in the Northern Territory of Australia, on the east coast of Arnhem Land. He heads the section "Types of Armed Conflict" and proceeds to state that "there are six distinct varieties of warfare among the Murngin" (p. 155): (1) a fight within the camp (2 deaths); (2) a secret method of killing (27 deaths); (3) a night attack in which the entire camp is surrounded (35 deaths); (4) a general open fight between at least two groups (3 deaths), in which both sides choose to fight at a predetermined location; (5) a pitched battle (29 deaths);

and (6) a ceremonial peace-making fight that is partly ritual ordeal (0 deaths). Warner recorded 72 engagements, over perhaps 20 years, and 96 killings. Almost all killings occurred in three of the types. In (2) men are attacked while sleeping. The deaths are said to be part of a feud. In (3) a night attack targets a specific person, but several of his relatives may also be slain. In (5) only two pitched battles were recorded in 20 years; a regional battle involving a large number of clans takes place. The two sides met halfway between their territories.

The three most lethal forms of armed combat (Warner calls conflict) appear to me to form a sequence in which there is escalation from a single killing to a battle in which many warriors die—(2), (3), and (5). The other three "varieties" appear to be outside the sequence—(1), (4), and (6). In my view we see in (2) and (3), versus (5), the basic pattern of warfare for uncentralized political systems, the ambush and line. Douglas Fry sees it differently for the Murngin. For Fry "with exceedingly rare exceptions" Australian armed combat "cannot be called war" (2006:159). For him the combat is "a class of *legal procedures . . . certainly not warfare*" (2006:159, italics in original). I believe Fry has misjudged: (1) warfare is turned in many cases into feuding, and (2) feuding is considered judicial, which permits him to conclude that the Australians are "unwarlike."

Our second example is familiar to us—the Yanomamö (see chapter 1). Chagnon (1997) describes five levels of armed combat that he says form a "graded system of violence." (1) Chest-pounding duels, (2) club fights, (3) spear fights, (4) raids, and (5) treacherous feasts. Since all five levels occur between sovereign villages, although club fights can occur within a village, they are seen by Chagnon as connected in that violence can escalate from one level to the next. I believe that it is usually the political community that is losing or behind in fatalities that chooses to move to a more lethal form of combat. This is similar to the feudists who try to even the score. The side that is behind attacks. A club fight within a village can lead to fissioning with the departing/losing faction forming a political community. The two political communities can now engage in warfare, meaning they can step back into the "graded system of violence."

The third and final example are "my people," the Higi (1968a). The Higi marriage system, which makes it possible for any man to attempt to obtain a wife who is married to another man, results in armed combat. The three types of armed combat—duels, staff fights, and feuds—which occur within the Higi political community, are classified as "internal conflict;" and the two types of armed combat—raids and line battles—which occur between this political community and other Higi political communities, are classified as "internal war." This distinction separates those types of armed combat that can be properly classified as warfare from other types that arise out of disputes generated within the political community. The raids, which result in ambushes, and the line battles form the basic pattern of warfare with which we are now becoming familiar. The three types of internal conflict stand alone in that a duel involves an individual trying to right a wrong (an attempt that may result in further loss); a staff fight is between villages within the polity (one reason being to select the new chief); and a feud is between two **patrilineages** (one of which is seeking revenge or compensation for a killing). W. E. A. van Beek, a Dutch ethnographer, describes the same dynamics for the neighboring Kapsiki (1987). This system creates a situation in which a constant state of enmity and hostility exists between political communities. War may break out at any time.

Each of the three examples presents a set of levels of armed combat that form a system different from the other two. The Murngin system consists of the basic pattern of warfare plus other types of combat that seem not to relate to the basic pattern. The Yanomamö system incorporates all five levels into an escalation sequence. The Higi types of armed combat belong to three separate domains. Feuding, in the minds of the Higi, is distinct from war between Higi political communities and the fights or duels with wooden weapons within the polity. This is further evidence that feuding needs to be kept separate from internal war. Each set of types of armed combat must be analyzed on its own terms to ascertain whether and how the types are connected.

Chapter Five

Foundations of War

To understand the foundations of war, I first examined armed combat between political communities and how it is conducted, that is, the military organization that goes on raids and engages in line battles and ambushes. This approach to the study of war is empirical, not inferential. I have reversed the usual direction of presentation in that I have taken what is known and worked backward. I do not believe that any single human trait or set of traits causes war or the basic pattern. Thus, given the basic pattern, we now consider what characteristics of Homo sapiens fit with this pattern. What biological and social characteristics of humans facilitate this basic pattern?

The *biological characteristics* of humans include anatomical traits and neuropsychological traits. The *anatomical traits* include: binocular vision; the ability to walk and run, and the ability to use shock and projectile weapons. The *neuropsychological traits* include: the amygdala, the primitive part of the brain that can trigger a fight or fear response; an impulse to respond aggressively to challenges; gratification in anticipation of taking revenge; and gratification in killing group members who are cheaters (Homo sapiens possess no inborn inhibition against killing). The *social characteristics* include: intragroup cooperation (kinship and local group), intergroup hostility, and socialization (primary—learning experienced early in life from caregivers—and secondary—learning that takes place throughout life as one encounters new groups of people). Straddling biological and social characteristics is the need to defend territory, which I discuss in terms of developmental stages.

BIOLOGICAL CHARACTERISTICS
THAT FACILITATE WAR

The warfare of Homo sapiens rests on or relies on both anatomical and neuropsychological traits. They are numerous, but I am not sure how many there are or whether one is more important than the others. I do not know whether they should be sorted into two separate categories, anatomical and neuropsychological, as I have done, but I know that both play a role in the basic pattern of war. I have clumped the traits into equal-sized groupings for ease of remembering. You may wish to add to or reorganize the list. You should do so if it helps with your understanding of warfare.

Anatomical Traits

Binocular vision inherited from our primate ancestors has been an indispensable aid to survival, hunting, and armed combat. With two eyes set apart, Homo sapiens have a horizontal field of view of approximately 200 degrees. Not only can we see ahead, we can see to either side. Binocular vision can give *stereopsis* in which *parallax* provided by the eyes' different positions on the head give precise depth perception. Our eyes can pinpoint the location of an object, animal, or human being, as well as estimate its distance from us. Such binocular vision is accompanied by *singleness of vision* or *binocular fusion*, in which a single image is seen despite each eye's having its own image of an object. Furthermore, two-eyed vision gives *binocular summation,* in which the ability to detect faint objects is enhanced. Such visual acuity as our ancestors had, paired with a projectile weapon, facilitate accurate targeting. Eye-hand coordination and an accompanying weapon is a lethal combination.

Two million years ago, upright posture and walking led to long-distance running. Homo erectus, the ancestor of Homo sapiens, had long, slender legs for greater strides. As the forests gave way to arid grasslands between two and three million years ago in East Africa, selective pressures turned the hominid residents into larger, taller, bigger-brained animals with the modern human body form. The gluteus maximus, the buttocks, stabilized the trunk and, in combination with long legs, made short- and long-distance running possible. With the emergence of archaic sapiens a half-million years ago and Homo sapiens by 100,000 years ago, the body form changed, becoming more slender and graceful. The change gave a survival advantage because it made possible escaping from many predators, running down prey, and transversing the savanna rapidly. The "new" body form probably made spear throwing easier; evidence for the use of throwing spears dates back 400,000 years ago to the area that is modern-day Germany. Many years ago

Alice Brues (1959) related body build to weapons' use: spearmen developed linear builds, while archers developed lateral builds and strong shoulders. Those who used blunt crushing implements developed heavily built physiques. Although warfare did not make the ideal warrior build, the physique that was developed through natural selection was ideal for warfare.

The hands of men and women through natural selection became ideal for armed combat. Although chimpanzees can throw objects and strike with sticks, they have anatomical limitations to tool using and tool making because chimpanzees have long fingers and a short thumb that tucks in along the side below the index finger. This combination creates a "power grip"—ideal for grasping branches and swinging through the trees, but not for throwing or thrusting objects. The hands of the upright-walking *Australopithecus afarensis* were similar to ours and, according to primatologists, could forcefully throw small round objects overhand (Marzke 1983). The opposing thumb creates a "precision grip" suitable not only for holding a rock or baseball but also for wielding a club or baseball bat. The small fingers against the thumb create a strong grip that makes possible a powerful downward blow, with club or ax, suitable for cracking coconuts or skulls (Otterbein 2004:47–50). The handle of a semiautomatic pistol is designed to take advantage of the gripping strength of the thumb against the three small fingers. Since we walk or run we do not need the power grip of an ape, which would permit us to swing through the trees. The precision grip is great for holding weapons, but war did not give it to us; instead, loss of tree cover and the development of an upright stance and walking probably did. Our vision and grip, plus the ability to run from an attacker or after an enemy, make us potential warriors if other conditions are present.

Neuropsychological Traits

Three neuropsychological traits facilitate war: the *amygdala,* an aggressive response, and the ability to interpret a situation. The brain of Homo sapiens has an ancient, almond-shaped neural structure called the amygdala that triggers a fight-or-flight response. The amygdala prompts releases of adrenaline and other hormones into the bloodstream. As part of the response, Homo sapiens may flex their arms and lean or angle away from adversaries. Neck and shoulder muscles will tense. The mouth becomes taut. The body may crouch. When you see this response in an adversary, duck or run like hell. You are about to be struck. If the adversary has a weapon you may have only one more second to live. Probation officers are instructed to face "clients" sideways for protection, and duelists, if right-handed and right-eye dominant, also stand sideways. Pistol shooters, if firing with one hand, do likewise. By turning sideways, the body becomes a smaller target.

The second neuropsychological feature is directly related to the amygdala. It is the aggressive response to a frustration or challenge. John Dollard and colleagues in 1939 formulated what became one of the most famous theories of the social sciences, the frustration-aggression hypothesis. They assume that:

> *aggression is always a consequence of frustration*. More specifically the proposition is that the occurrence of aggressive behavior always presupposes the existence of frustration and, contrariwise, that the existence of frustration always leads to some form of aggression. (Dollard et al. 1939:1; italics in original)

More recently, a team comprising a psychologist and geneticists put forth a paradigm that is frequently observed in mammal groups: "Challenge elicits aggressive response." The response is an attempt to reassert status within the group, to gain access to females, or to obtain food and shelter. The team also concluded that another variety of human aggression is predation. Mammals are hunted for food, they also hunt animals that prey on them and are major competitors for food (Blanchard, Hebert, and Blanchard 1999).

Interpretation plays a major role in whether an aggressive response is elicited. The individual must judge the intent of the other person or animal. Thus, worldview becomes important; that is, if a group of individuals view the world as hostile and threatening, they may interpret the actions of others as a challenge or an attempt to frustrate their progress toward a goal. The hypothesis or paradigm needs elaboration: an action viewed as a threat or a challenge from a hostile world leads to a counterattack.

Our third neuropsychological feature takes us a step further. Homo sapiens achieve satisfaction in blocking the threat and injuring or killing the attacker. Researchers have found that a subcortical region of the brain, called the *dorsal striatum*, increased its consumption of oxygen (that is, was "activated") when punishing an aggressor or defector (a norm violator). The punishing response that triggers feeling good about getting revenge blocks or supersedes feeling bad about having been violated (Knutson 2004:1246). This satisfaction response happens even when the punisher places himself or herself in a situation of personal cost, and the striatum is even activated in anticipation of satisfaction. Researchers further found that the dorsal striatum becomes activated with the "altruistic punishment" of "defectors" who had abused the trust of the community (deQuervain et al. 2004). The implications of this research are far ranging for the interests of this book.

The activation of the striatum does not create homicide, feuding, war, or capital punishment, but it helps explain why there are eager participants in such violent activities. It explains:

- Why we enjoy seeing the bad guy get his just deserts—punishment in line with the wrong committed. For example, if a bully falls to his death, a common theme in movies, "he got what he deserved."

- Why some individuals enjoy killing, for example Jack Ganzhorn, gambler, gunfighter, and U.S. soldier in the Philippine War, who late in life wrote *I've Killed Men* (1959). Frequent paragraphs in the book describe Ganzhorn's pleasure in killing.

- Why we want to take revenge, such as the attempt at a counterkilling in a feud, even though we put ourselves at great risk.

- Why we initiate raids in a military situation, even though the participants, the warriors or soldiers, know that some, even all, may be killed.

- Why capital punishment is universal. We, at least some Homo sapiens, enjoy killing bad guys and many more enjoy watching executions.

SOCIAL CHARACTERISTICS
THAT FACILITATE WAR

Two characteristics of Homo sapiens are of extreme importance. They can be seen as different or as two sides of the same coin. They are cooperation within the group and hostility and fighting between groups.

Intra- and Intergroup Behaviors

Intragroup cooperation played a vital role in the survival and evolution of early hominids. It probably began in the family of the partners and their offspring and extended to related individuals that formed the local group. For the families and the local group to survive in a hostile environment their members had to cooperate. If they did not, they did not survive. Individual survival depended on group survival. I believe those early hominids cooperated using a rudimentary language to communicate. "Cooperation would be useful not only for defending against predators, whether they be carnivores or other early hominids [not necessarily of the same species], but also for hunting other animals" (Otterbein 2004:40).

In the 1960s studies of Fraternal Interest Groups showed that kinship groups protect individual members who in turn protect the local group (Otterbein and Otterbein 1965). A study later in the decade showed that a polity whose military organization is more efficient and effective than its neighbor's would survive and expand territorially (Otterbein 1970). In the Foreword to *The Evolution of War,* anthropologist Robert Carneiro states:

> When societies fight, the cultural equivalent of natural selection
> comes into play. This "cultural selection" operates in two ways:
> intrasocietally and intersocietally. . . . In intersocietal selection . . .
> the unit on which selection operates is not the culture trait . . . , but
> the society bearing it. (Carneiro 1970:xii)

By 1975 zoologist Robert Bigelow was arguing that group selection was
more important than individual selection (Bigelow 1975). Science
writer Bruce Bower has traced the development of what is now known
as "cultural group selection" from its 1960s roots (1995).

What has come to be an important element of the theory of cultural
group selection is the recognition that groups often punish or kill their
own, their so-called "cheaters" (Vogel 2004). The biological basis for
being able to do this has been described previously. My cross-cultural
study of capital punishment showed that executions occurred more frequently and for more reasons than most scholars have recognized (1986).
I found that even small-scale societies, including hunting-gathering
bands, executed individuals who disrupted the social order. These individuals were usually recidivist killers, witches, violators of sexual or
religious norms (see "Capital Punishment" in chapter 4). I later argued
that this tended to remove from the gene pool of the group, genes for
undesirable behavior (1988, 2004).

Defense of one's group from a neighboring raiding party, and the
revenge that is taken for the raid and any resultant injury or fatality,
can create *intergroup hostility* that may last for centuries (Boehm
1984). The Bellona described in chapter 4 are an example. A cause of
war may be retaliation. Some scholars have recommended a "deterrence strategy," others a "tit-for-tat strategy" for deterring or curtailing
war. I do not see how these strategies can ever lead to the end of hostilities. (I critique this notion in chapter 8.) War can begin with and
spread from attacks and counterattacks of polities in a setting referred
to by archaeologist Lawrence Keely as "bad neighborhoods" or "raiding
clusters" (1996:127–128). "The aggressive societies at the heart of these
raiding clusters were rotten apples that spoiled their regional barrels."
Keely points out that "evidently, then, one factor intensifying warfare
is an aggressive neighbor. Most societies that are frequently attacked
not only fight to defend themselves, but also retaliate with attacks of
their own" (p. 128). However, not all attacked societies retaliate.

In previous publications I have underestimated the number of
peaceful societies. In a cross-cultural study of warfare, I found only 4
out of 50 societies that lacked both military organizations and war; all
four were isolated, essentially without neighbors (1970). A recent
examination of the frequency and type of war of the 46 warring societies in this sample led to the discovery of two more that were attacked
(defensive external war), but did not retaliate (engage in offensive
external war), and a third that I classified as having a military organi-

zation, but did not engage in internal war or attack culturally different people, and itself was infrequently attacked. The societies attacked that did not retaliate were the Pueblo of Santa Ana in the Southwest and the Trumai, a Central Brazilian tribe in a refuge region. The non-warring Monache were a California hunting-gathering band. Their military organizations consisted of male hunters. These are societies that would have been peaceful if isolated.

Thus, early groups of Homo sapiens formed polities of cooperating individuals who would have come to see themselves as members of their group. When they encountered another polity, they met individuals who regarded themselves as being members of their own group. If the "meeting" turned violent, the revenge motive could set into motion killings and counterkillings. Friends and enemies emerge. The individuals in your polity are friends, the individuals in the other polity are enemies.

Ethnocentrism

What has just been described is known as ***ethnocentrism.*** The term was coined by sociologist William Graham Sumner. His discussion of ethnocentrism in *Folkways* is probably the most famous passage in the social sciences. Because of its classic stature and its relevance to the study of warfare, the entire set of passages will be quoted (1906:12–13):

13. THE CONCEPT OF "PRIMITIVE SOCIETY"; WE-GROUP AND OTHERS-GROUP. The concept of "primitive society" which we ought to form is that of small groups scattered over a territory. The size of the groups is determined by the conditions of the struggle for existence. The internal organization of each group corresponds to its size. A group of groups may have some relation to each other (kin, neighborhood alliance, connubium and commercium) which draws them together and differentiates them from others. Thus a differentiation arises between ourselves, the we-group, or in-group, and everybody else, or the others-groups, out-groups. The insiders in a we-group are in a relation of peace, order, law, government, and industry, to each other. Their relation to all outsiders, or others-groups, is one of war and plunder, except so far as agreements have modified it. If a group is exogamic, the women in it were born abroad somewhere. Other foreigners who might be found in it are adopted persons, guest friends, and slaves.

14. SENTIMENTS IN THE IN-GROUP AND TOWARDS THE OUT-GROUP. The relation of comradeship and peace in the we-group and that of hostility and war towards others-groups are cor-relative to each other. The exigencies of war with outsiders are what make peace inside, lest internal discord should weaken the we-group for war. These exigencies also make government and law in the in-group, in order to prevent quarrels and enforce discipline. Thus war and peace have reacted on each other and developed each other. One within the group the other in the intergroup relation.

The closer the neighbors, and the stronger they are the intenser is the warfare, and then the intenser is the internal organization and discipline of each. Sentiments are produced to correspond. Loyalty to the group, sacrifice for it, hatred and contempt for outsiders, brotherhood within, warlikeness without,—all grow together, common products of the same situation. These relations and sentiments constitute asocial philosophy. It is sanctified by connection with religion. Men of an others-group are outsiders with whose ancestors the ancestors of the we-group wage war. The ghosts of the latter will see with pleasure their descendants keep up the fight, and will help them. Virtue consists in killing, plundering, and enslaving outsiders.

15. ETHNOCENTRISM is the technical name for this view of things in which one's own group is the center of everything, and all others are scaled and rated with reference to it.

The concept of ethnocentrism, and the explanation behind it, is one of the most fruitful lines of research in anthropology, political science, social psychology, and sociology. It has given rise to the ***conflict/ cohesion hypothesis*** (Coser 1956). When there is conflict between groups, there is cooperation within the groups in conflict. A famous field study of the conflict/cohesion hypothesis was the Robbers Cave experiment conducted by social psychologist Muzafer Sherif and colleagues. At a boys' camp, the campers were divided into two groups called the Eagles and the Rattlers. They resided in separate areas. As the groups developed their own identities, "garbage" fights broke out in the cafeteria, along with raiding and counterraiding of the camp sites (Sherif et al. 1961). Highly negative terms were applied to members of the other group. The counselors (actually researchers) brought the experiment to a close by introducing Superordinate Goals. Tasks were created, such as a truck that would not start, and many boys were required to move it. Both groups cooperated in pulling it out of the mud (Sherif et al. 1961).

Socialization for War

In a climate of hostility between political communities men and women grow up with positive views of their neighbors and other members of their political community and negative views of the people in other political communities, even if the polities are within the same culture. "The origin of ethnocentrism lies . . . in conflict between early hunting/gathering bands" (Otterbein 2005b:xv). Ethnocentrism is a normal aspect of socialization, especially socialization for war. These attitudes—both positive and negative—are learned, informally and formally, and lead the individual to support the military organization, if one exists. If war occurs or is likely to occur between these polities, military organizations will need to be developed to ensure polity survival. Any polity with a military organization needs to recruit young males, and sometimes women, into the military. After the appropriate attitudes have been learned, the most effective means of recruiting appears to be the training of young men and women—informally and formally—in the use of weapons.

Ethnographic examples are numerous. Dani youth engage in battles with toy spears. There are scenes of this in *Dead Birds*. Yanomamö youth have "battles" with bows and arrows in the plaza in the center of their village. Scenes are in several of Chagnon's films. The cover of my selected works, *Feuding and Warfare*, has a photograph of two Higi boys holding bows and carrying arrows and knives (1994c). It is a posed picture—I had asked them to show me weapons. However, one boy, without a request from me, assumed a crouching position of a warrior on the battlefield. I also saw and photographed young boys with small bows target practicing; the targets were upright stalks.

Although shooting weapons may be a recreational sport and practice for hunting, the skills learned are directly transferable to the battlefield. Many American war heroes over the past century were deadly marksmen: Jack Ganzhorn from the Philippine War killed many insurgents (natives who did not want the U.S. military there). Alvin York from World War I had been a conscientious objector, but once sent to France, he killed many Germans with both a bolt action rifle and a .45 semiautomatic pistol. Audie Murphy in World War II killed many Germans in close combat with an M1 carbine. Anthony Herbert in Vietnam killed many Vietcong with the M16 rifle. Carlos Hathcock, Marine Sniper, was described in chapter 2. In their autobiographies, or, in the case of Hathcock, biography, these men all described hunting experiences they had in their lives. Fighter pilots, such as Robert Scott, have also described hunting in their childhood. The pilots, however, typically used shotguns to shoot birds on the flight. Learning to shoot a moving target, needless to say, is a big help to a fighter pilot headed toward acehood.

I do not claim that skill with weapons or firearms makes a person want to be a warrior or soldier, or necessarily turns a nation of hunters into soldiers. But I am saying that those who participate in the shooting sports make good soldiers. Many of the earliest and best rifles were made in Germany, and German soldiers became the best. However, large numbers of Germans from the time of the Reformation were pacifists. For this reason, William Penn, an English pacifist, recruited them in the 18th century for his colony of Pennsylvania. Yet, from Pennsylvania to North Carolina they made the famous Kentucky rifle, which they sold to the Scots-Irish (Otterbein 2000b:240). These rifles were then used in feuds until replaced with Winchester repeating rifles.

Another important informal source for socialization for war is combative sports. Some are individual competitions such as boxing, wrestling, and the martial arts. Others are team sports such as football, lacrosse (Iroquois), or the Central American Indian ball game played on a special ball court (Aztec to Zapotec) (Marcus and Flannery 1996). Anthropologist Richard Sipes, in a cross-cultural study, demonstrated that offensive external war and combative sports are directly related (1973). They are related because both derive from sociocultural selection for warlike societies. War makes combative sports a useful tool of military socialization, and such socialization contributes to military efficiency.

A third informal source is driving or riding horses. If a military organization employs chariots or cavalry, trained drivers or riders are essential. There is not time after one is conscripted to learn these skills. If the society is **nomadic,** the skills are known by all. In agricultural societies the better-off families are likely to have horses. The Confederate States army had an excellent cavalry, since plantation owners had grown up in the saddle. Horsemanship until the 20th century was such an important skill it was essential for officers and the rapid attack units of an army, known as the cavalry.

Once in the military, socialization becomes formal. Practice in performing maneuvers becomes important. Our earliest examples of war art, cave paintings of marching warriors and monumental relief sculptures of battles such as Rameses III defeating the Sea People, suggest that drills took place from the earliest periods; from works created nearly 5,000 years ago in Mesopotamia, we "see" marching troops and phalanxes. Practice took place with military weapons, including catapults and cannons.

Attitudes toward War

In addition to skills useful in war, attitudes toward war are an important aspect of socialization for war. Societies that frequently engage in war and believe that success in war is essential to their existence develop attitudes that are congruent with this belief. I begin my class on warfare with a question that is designed to elicit class responses: "Why do we (meaning the United States) glorify battlefields?" The answers differ each year, although they have much in common. The question presupposes that the U.S. does glorify battlefields. Honoring the dead is always on the list. You, the reader, can make your own list. I believe that you will slowly begin to realize that the U.S. is a militaristic society. Values of patriotism are inculcated into American youth whether native born, naturalized citizens, or illegal aliens (some willing to serve in the U.S. military for a promise of future citizenship). Visiting battlefields seems to inculcate martial values. Youth learn about the battle and see statues of the war heroes—they are usually mounted on horses, not lying dead on the battlefield.

Similar to the battlefields are historic reconstructed villages. Often military reenactors in Revolutionary War or Civil War uniforms parade with their muskets and rifles. They fire cannons and stage mock battles. The bands play. It is very exciting. Old forts seem to be everywhere. The personnel are usually dressed in appropriate costumes. Historic tourism is on the increase. I could not begin to visit all the battlefields, forts, or outposts that there are in the U.S.

The warring states of prehistory that were on the threshold of developing written languages, such as cuneiform or hieroglyphics, have left a record of wall paintings of warriors, sculptures, and monuments. Some states, like the Zapotec and coastal peoples of Peru, made stone monuments that showed disembowelment and dying warriors. Archaeologists may interpret these as intended to frighten the enemy (Marcus and Flannery 1996). I think it was just as likely that they were intended to encourage young warriors and would-be warriors to perform these dastardly acts on the enemy. Torture arises out of ethnocentrism. Communications expert David Perlmutter has reviewed some of these in *Visions of War: Picturing Warfare from the Stone Age to the Cyber Age* (1999). I believe that in these early warring states the evidence of ethnocentrism is strong and that youth are being encouraged to willingly participate in combat.

DEVELOPMENT OF TERRITORIALITY

Territoriality, the need to defend territory, is seen by many biological scientists as having a genetic basis, and perhaps as the origin of war. On the other hand, most anthropologists and other social scientists view territoriality as a cultural construction, not a genetically based characteristic. Notions of territory vary from one culture to the next. Playwright/author Robert Ardrey popularized this genetic point of view in the 1960s in a series of books, one aptly titled *The Territorial Imperative* (1966). Merely citing the book at that time evoked the ire of many cultural anthropologists. I side with my brethren, but my ire is not aroused.

I think that territoriality is incidental for many, if not most, ***hunting and gathering*** bands. It is their territory because they are hunting in it; boundaries do not seem to be fixed and known to them. I am calling this the first stage in the development of territoriality. If hunter-gatherers move frequently, they are not likely to develop notions of fixed boundaries. Yet, the campsites and their households will be treated as the exclusive property of the inhabitants and defended from animal and human attackers. If campsites of two or more bands are close and form a Macroband, meaning they are members of a single political community, the land between the sites will probably be seen as the people's, and the land will be defended from raiders. (More information about Macrobands is in the next chapter.)

Layton and Barton have argued that "variation in the behavior of hunter/gatherers in different environments suggests human territorial behavior evolved in response to particular ecological stresses" (2001:13). Specifically, those groups that have developed a "regional community" are more likely to be occupying an environment with dense and predictable resources. These areas are more likely to have fixed boundaries and be defended. "The common principle may apply that it is the populations living at high densities that are prone to boundary defense and its corollary, cross-boundary raiding" (2001:19). Populations living in sparsely populated areas with resource scarcity are unlikely to have fixed boundaries, and they allow members of other bands to pass through or they attack and chase off intruders when encountered. This appears to be true for four of the eight types of bands described in chapter 6. They are the bands that rarely engage in warfare. The Australian Aborigines appear to be an exception. The fate of a hunter that wanders into a territory with fixed boundaries is likely to be death. Raymond Kelly (2000) has described this for the Andaman Islanders, who shoot intruders on sight, the "shoot-on-sight" tactic described in chapter 2.

The second stage arises in those regions where resource competition has become intense. If it is dangerous to be hunting near the edge of the range of one's group, boundaries will come to be set back, away from one's enemies. Presumably both warring peoples will pull back, thus creating a buffer zone. In such situations a buffer zone may arise because hunters fear their enemy may be hunting near them. Buffer zones may be two to three miles wide and become refuge areas for animals and birds, who then are not hunted. Examples from ethnography are known. The red deer of Virginia survived in a buffer zone (Hickerson 1965). Recent reports from the DMZ between North and South Korea have asserted that rare animals and birds survive there.

In the third stage, once a territory has become valuable to the people, one group may "see" boundaries where the other group does or does not. These are natural boundaries like rivers, lakes, and deep ravines that are difficult to cross and easy to defend. In what has been referred to as "the first scientific account of an Indian tribe," Lewis Henry Morgan (1818–1881) devotes an entire chapter to Iroquois place names and the paths that linked them to rivers and lakes (1962:v, 412–443[1851]). Since it was the 19th century, Morgan and his informants used New York State maps. In a fourth stage, or perhaps at the same time, a local sage, philosopher, or political leader will be able to make a primitive map that will show important locations, real and mythical, along with rivers and bodies of water. The distances shown on the map may be greatly inaccurate.

The development of territoriality in the West is stage five. The state, as we know it, developed in the 15th century and so did fixed territorial boundaries. Warfare, fortifications, cartography—maps showing locations of forts, roads, villages, cities, bodies of water, and even the locations of battles—went hand in hand with the development of national boundaries (Virga and the Library of Congress 2007:148). Many of the famous sites of ancient battles are unknown because there were no maps. Examples abound: Boudicca's Uprising against the Romans in the British Isles, AD 61 (Southwaite 1984:22), and the Chou defeat at the Battle of Maye of the Shang Dynasty in China, 1045 BCE, in one of the great chariot battles (Otterbein 2004:159–169).

Chapter Six

Origins of War
Two Paths

Warfare developed along two separate paths. The hunting of large game animals was critical to the development of the first path. Early hunters, working as a group in pursuit of game, sometimes engaged in attacks on members of competing groups of hunters; they devised a mode of warfare based on ambushes and lines.

At the origin of the second path were foragers who did little hunting but depended largely on gathering for subsistence, became sedentary, and domesticated plants. Intergroup aggression was absent among these early agriculturalists. The first states developed only in these regions, but once city-states arose, a mode of warfare based on line battles and siege operations sprang forth.

HUNTER-GATHERER WARFARE

Early Humans

Whether warfare arose early in our evolution, perhaps eight million years ago, only in the late Upper Paleolithic (about 20,000 years ago), or even not until after the origin of agriculture, is probably the most controversial topic in the Anthropology of War. On the side of the antiquity of war are most primatologists and some archaeologists; on the side of the recent emergence of war are some archaeologists and many cultural anthropologists. Advocates for the early warfare position focus on the behavior of chimpanzees. They project back to what is known as the "evolutionary platform" from which our ancestors' behavior evolved; that is, to the time, between five and eight million years ago, when the pongid line split from the hominid line (King 2008:6). Chimpanzee homicide is called "coalitionary killing" and so is the fighting of our non-Homo sapiens ancestors (Wrangham 1999). But what we know about the latter is *nothing*.

I have not let lack of knowledge stop me. In a 21st-century tome on the origins of war I have a section titled "Speculations about Early Warfare" (2004:60–62). To broadly generalize from "Speculations," it is sufficient to say that I believe that during the time span from Homo habilis to Homo erectus, clubs were probably used by cooperating groups of hunter-gatherers, sometimes with Fraternal Interest Groups present, in ambushes and line "battles." By 400,000 years ago, archaic Homo sapiens were using heavy tapered throwing spears to hunt wild horses in central Europe. Warfare *could* have occurred at this time. ("Could" appears frequently in "Speculations.")

Modern Homo sapiens evolved from archaic Homo sapiens between 100,000 and 200,000 years ago in Africa, and Homo neanderthalensis evolved over a 400,000-year period in Europe, also from archaic Homo sapiens. Both of these homo species moved into the Middle East and occupied land in the Levant. Both species used stone points as spear tips on thrusting spears and on throwing spears. This increased hunting efficiency and could have made the killing of fellow species members easier. I concluded: "While warfare was probably rare until about 40,000 years ago, it did occur on occasion. Fraternal Interest Groups, weapons, and hunting form a complex—given its ancient origin it can be called the eternal triangle" (2004:62).

In my story, shortly before 40,000 years ago, the Modern Homo sapiens in North Africa invented the spear thrower, a short hooked stick held in the hand, which could launch a small spear or dart with a small stone point. The spear thrower increases throwing distance (see figure 2.1). The spear thrower, called an atlatl after the Aztec name for this weapon, was used to hunt animals that made a target large enough to be struck by a dart, such as a gazelle or a small antelope like a duiker, but not so large that the animal posed a danger to the hunter. When the atlatl got into the hands of the sapiens along the Nile River, they were able to advance farther into the Middle East. Prior to this time, neither the Homo sapiens nor the Neanderthals appear to have been able to dominate in the area (Shea 1997, 2006). I believe the greater throwing range of the atlatl caused the Neanderthals to retreat. Eventually, the Neanderthals occupied several separate refuge areas in Europe, and then died out (Finlayson 2004). As Homo sapiens advanced, Homo erectus groups that occupied eastern and southern Africa and most of Asia including China probably retreated to isolated islands like Java and Flores, and then died out about 20,000 years ago. On Flores they became smaller and developed into a new species, Homo floresiensis (Shipman 2006). Homo sapiens rapidly spread over Asia, then entered the Americas from about 18,000 to about 26,000 years ago. At this time "a decrease in the sea level exposed a large land mass joining the northeast of Asia with the northwest of America called Beringia." Between about 12,000 and 18,000 years ago "the Beringia landmass was being reduced due to sea level rise" (González-José et al. 2008:184). I believe that people from Beringia proceeded rapidly down the west coast of North and South America, reaching the southern west coast of the latter continent by about 14,000 years ago. This is referred to as the "Out of Beringia" or "Single Wave" model (González-José et al. 2008:185). No ape or hominid species was encountered. Few scientists give any credence to the idea that a large ape referred to as "big foot" could have lived then or now in Canada or in the western United States. Homo sapiens spread inland on both continents reaching even the eastern part of the North American continent (today Pennsylvania) by about 12,000 years ago. Warfare arose.

The evidence for warfare is scant. However, as archaeologists come to inherit the earth, I believe more evidence will be found. When I was a student, nothing had been found to support the idea that warfare occurred in the Upper Paleolithic. Today, archaeologists have found cave and rock wall art that depicts the killing of individuals and shows battle scenes. Human bones with projectile points lodged in them and group burials of individuals who had been killed violently complete the evidence for warfare in the late Upper Paleolithic (Guilaine and Zammit 2005:40–81[2001]; Otterbein 2004:71–77).

I generalized even further. The eternal triangle began weakly. Hunting, Fraternal Interest Groups based on virilocal or patrilocal residence, and weapons slowly formed a complex. Hunting led to virilocality. A cross-cultural study using a sample of hunter-gatherers has shown that extensive hunting and fishing are related to virilocal residence and hence the formation of Fraternal Interest Groups (Otterbein 2005a). Another cross-cultural study using a similar sample of hunter-gatherers shows that hunting, particularly of large game, and high warfare frequency are related (Otterbein 2004:85–90). This led to the generalization that as hunting increased so did warfare; as hunting decreased, with the scarcity or extinction of animals, so did warfare. As the African Homo sapiens expanded into Asia and Europe the large animals came to be overhunted. This led to a decrease in hunting and also a decrease in warfare. With fewer large animals to hunt and greater reliance on small game and seeds and plants, people became more sedentary. With a sedentary life and less reliance on weapons for hunting, there was little warfare between hunting and gathering bands. Our examination of eight hunter-gatherer types in the next section will show that this relationship has held into recent centuries. The sedentary groups domesticated plants in the absence of warfare, a situation that brought forth the development of political complexity. The second path to war then began, and another story occupies our attention.

Types of Hunter-Gatherers

Two tasks lie before us. The first task is to develop a typology of hunter-gatherer societies, and the second is to establish which types are dominated by groups with warfare. Hunter-gatherer cultures in the ethnographic record will be classified as one of eight types (Otterbein 2008). I will not be coding a sample of hunter-gatherer societies as to which have warfare and which do not; coding from ethnographic descriptions is a procedure followed by researchers conducting a cross-cultural study. I will be classifying the types as to whether the majority of cultures within each type has or does not have warfare. I recognize that the proper methodological procedure is to derive types from actual cases, and indeed I have described some specific peoples.

I began the typology with George P. Murdock's listing of hunter-gatherer bands that he created for the Man the Hunter Conference of 1966, which was attended by 50 anthropologists and was held to examine the status of the world's hunter-gatherers (1968a, 1968b). Murdock's approach was to sort all the known hunter-gatherers by geographic region, 27 regions in all. He proceeded to "set aside" several distinct types. He first set aside the Mounted Hunters, also called Equestrians (C. Ember 1978). Examples are Plains Indians, like the Cheyenne and Sioux, who acquired horses indirectly from the Spaniards, mounted up to hunt bison, then waged war against each other and, finally, against the U.S. Cavalry.

A second type that Murdock set aside are the Settled Fishermen of the Northwest Coast of North America, from Alaska to Washington State. They have been described as Complex Hunter-Gatherers. They had a complex social and political organization, usually referred to as a chiefdom. These are peoples like the Kwakiutl, who lived in large wooden houses in coastal villages. They are famous for their totem poles. They built ocean-going dugout canoes propelled by many paddlers. From their canoes they fished and hunted sea mammals and raided neighboring villages. Salmon runs gave them bountiful harvests. They were virilocal and waged war ruthlessly (Benedict 1934; Codere 1950).

> Where, however, the bellicose chief of a large settlement was able to invite or compel the assistance of his weaker neighbours, considerable war fleets occasionally assembled and attacked villages fifty to one hundred miles away. Although small quantities of booty, such as sea-otter skins and the heads of slain enemies, were brought back from a successful attack, the humbling of the enemy and the destruction of his village appears always to have been the main objective. . . . Attacks were made before dawn, and the crews

of the war canoes were usually carefully organized, each crew being assigned to a particular house in the enemy's settlement. Some bore the brunt of the fighting, others carried brands to fire the houses and ropes for binding captives to be carried off as slaves. (Forde 1963:95–94)

These two categories of hunter-gatherers have been well-known to anthropologists and their students for over a hundred years.

A third type set aside by Murdock are Incipient Tillers. This category seems largely unknown to other anthropologists. Tillers are horticulturalists and Incipient Tillers are Settled Gatherers who plant seeds. Murdock (1968a) identifies one example, the Gě who reside on the savanna of central South America. Incipient Tillers may be gatherers who have settled and may domesticate plants and later animals, or they may be hunter-gatherers living on the periphery of horticultural or agricultural peoples, who furnished the seeds to them. This type would have occurred from the **Neolithic** onward. Another group that appears to fit this category is the Semai, a non-Negrito Malay population Robert Dentan believes were hunter-gatherers who became horticulturalists (1979). The Semai hunt with the blowpipe and are famous as a people without warfare. The Semai are culturally similar to their neighbors, the Semang, a nomadic hunter-gatherer Negrito people who hunt with the bow and arrow like the similar Andamanese (both the Semang and Andamanese are categorized as Simple Hunter-Gatherers, not Incipient Tillers).

Another culture that fits the Incipient Tillers category are the extremely warlike Waoranai of northwest South America. Their ethnographers, Clayton and Carol Robarchek, previously did field research with the Semai, and conducted a **controlled comparison.** They deliberately chose for their second study a people they regarded as being until recently "the most violent society on earth"; "60% of Waoranai deaths have been homicides" (1998:10, 19). Their subsistence is based on hunting, with blowpipes and nine-foot spears, and swidden gardening (1998). Their physical environment is similar to that of the Semai; however, their culture is radically different (Robarchek and Robarchek 1992). The Semai reacted nonviolently when they were regularly raided by Malay; the Waoranai, on the other hand, kept out foreigners by killing them. The first missionary they met, they killed; five Americans were speared to death in 1956.

After examining the remaining regions and their hunter-gather peoples, Murdock decided that the Australian Aborigines were so different from all the others that he pulled them out, thus creating a fourth category. His reason was that they were virilocal, practiced polygyny, had unilineal descent (patrilineal) and complicated kinship systems, as well as warfare. Murdock's large residual category, after four categories had been set aside, has come to be called Simple

Hunter-Gatherers. They are usually **multilocal,** monogamous, **bilateral,** bands composed of extended families, and they did not engage in warfare. The Andaman Islanders are considered to be a major example, although they engaged in both internal and external war (Kelly 2000). Approximately one-third of Simple Hunter-Gatherers do engage in war (Fry 2006). I further draw out from Murdock's fifth category two more distinct types, then subdivide one of these types into two types. Five plus three equals eight.

For the next category I pull out the Symbiotics. This is my term, but the category has long been recognized. Some hunter-gatherers are in a symbiotic relationship with a larger, more powerful agricultural society. They are on the periphery, geographically, of that society. The relationship is symbiotic because their neighbors provide produce the farmers have grown, and in turn, the hunter-gatherers provide their neighbors with products of the forest, such as wild game. The most famous example of a Symbiotic are the Mbuti of the Ituri Forest of the Congo Basin, a people short in stature, hence called pygmies, who hunt in one part of their range with bows and arrows and in another part with nets (Service 1979). The Mbuti live in the interior forests. Some hunt elephants with a short stabbing spear. Some surround a single elephant and attack from several sides. Patrick Putnam's description of a lone hunter crawling under an elephant at a water hole and driving the spear into the animal's abdomen ranks as one of the all-time great hunting stories (1948). The entire band partakes of the flesh.

My reason for identifying a Symbiotic category is that it includes groups who have a complex subsistence base, which consists of what they hunt and gather plus agricultural products. Their diet only partly reflects their subsistence activities, which are heavily dependent on hunting. Agricultural products are obtained through trade. For example surplus meat is exchanged for corn beer and handicrafts. To me, the Symbiotics do not look like Simple Hunter-Gatherers. They are locked in a dependent relationship with their neighbors. Indeed the Pygmies have had such long contact with their neighbors that they speak different Bantu and Sudanic languages, depending on who their neighbors are. They do not have their own language (Barnard 2004). Under these circumstances I would not expect them to use their hunting weapons in war unless they are hired as scouts or mercenaries.

The next category, the Big Game Hunters, easily subdivides into Macrobands and Microbands. They are frequently spoken of in descriptions of the Upper Paleolithic in the Old World and the Archaic in the New World. Perhaps best known are the Reindeer Hunters of northern Eurasia and the Clovis Hunters of mammoths in the Great Plains of North America. *Prehistoric Hunters of the High Plains,* by archaeologist George Frison (1978), covers both the archaeology of the region and experiments with hunting weapons. In his conclusion he also critiques the findings of the Man the Hunter conference that hunter-gatherers were primarily gatherers (1978). Frison concludes that their subsistence came largely from hunting. One hunting technique involved a large number of men, and perhaps women, surrounding a group of frightened animals. It required a Macroband, composed of perhaps three Microbands, to provide enough hunters to successfully hunt so many animals at once or to take on a huge dangerous animal. If a single animal like a moose is being stalked, only one to three hunters will be needed (Coon 1971). What may have helped tie the Microbands together was kinship, resulting from the marriage of men and women to persons in the other Microbands. The Macroband was likely to be virilocal, the Microbands multilocal.

The division of Big Game Hunters into Macro- and Microbands was suggested by George Frison in *Survival by Hunting* (2004:224–225). In his conclusion he says that hunts of a single animal by two men (two men likely to be more successful than one hunter) from one band composed of about five nuclear families gave way at times to communal hunts by three bands. Communal hunts required many more people than one band could supply. According to Frison, a Macroband would kill and eat about 200 large animals in a year. If a Microband numbered about 25 persons, then three combined Microbands would yield a Macroband of about 75 persons, sufficient to create a force of about 20 hunters; this group could also serve as warriors if their band was attacked or if they had reason to attack another band. The switch to communal hunts occurred in the Great Basin of North America about 5,000 years ago. With the decline of large animals in both North America and northern Eurasia, communal hunts began to occur with less frequency. My reason for dividing Big Game Hunters into Micro- and Macrobands is that the former are unlikely to engage in warfare (although self-help and feuding could occur), while the latter are likely, with their substantial hunting body, to engage in warfare from time to time. Defending a kill site where animals can be cornered or driven off a cliff is a likely location for a battle.

An example of a Microband would be the Cree of northern Canada (Bishop 2007). An example of a Macroband would be the reindeer-hunting Yukaghir (Forde 1963) or the early Reindeer Tungus (herders) of Siberia before the reindeer became semidomesticated (Forde 1963; Service 1963). Yukaghir internal war seldom occurred, but the Yukaghir and Tungus raided each other and captives were taken. The Reindeer Tungus (herders) appear to have a relationship to the Yukaghir (hunters) similar to the relationship of the Semai (Incipient Tillers) to the Semang (Simple Hunter-Gatherers). Both the Tungus and the Semai were once, in prehistoric times, hunters like the Yukaghir and the Semang. But over time, they hunted less and domesticated plants (Semai) and reindeer (Tungus).

Which of the eight categories are likely to include cultures that engaged in warfare, and which do not? Which of the eight categories are likely to have peoples that once lived in the Upper Paleolithic, and which societies are of more recent origin? Figure 6.1 shows the answers to both questions.

	Type A Societies: Warfare Frequent	Type B Societies No/Infrequent Warfare
Upper Paleolithic	Big Game Hunters, Macrobands Australians Settled Fishermen	Simple Hunter-Gatherers or Foragers
Recent	Mounted Hunters	Big Game Hunters, Microbands Symbiotics Incipient Tillers

Figure 6.1   A Typology of Hunter-Gatherer Societies

The figure makes it clear. Some early peoples might have engaged in warfare and some probably did not. In the Upper Paleolithic, it was the hunters who had warfare and the gatherers who did not. It is these groups that form the origins of the two paths to war (see frontispiece). When the number of large wild animals declined, some hunter-gatherers became partially dependent on agricultural products from their neighbors. Indeed, it seems to me that there is a developmental sequence from Foragers to Symbiotics to Incipient Tillers. Where the number of wild animals did not decline, the hunting people who, in recent times, had obtained horses continued to hunt and wage war.

Do Hunter-Gatherers Have War?

Since the Man the Hunter conference, the trend and tendency in hunter-gatherer studies has been to homogenize hunter-gatherers so that they form a single category of similar peoples (Barnard 2004). One culture has served as the model or type, the !Kung of the Kalahari Desert who had been an isolated group of San, formerly called Bushmen. The !Kung have replaced the Aranda and other Australians as the quintessential hunter-gatherer society. Hunter-gatherers today are seen as multilocal (not virilocal), and bilateral (without unilineal descent groups); polygyny is limited (often limited to sororal polygyny); the nuclear family dominates (but extended families frequently occur); women as gatherers bring into the camp the bulk of their subsistence; *and* warfare is rare. One well-known ethnography on the !Kung is titled *The Harmless People* (Thomas 1959). These people have been in such long contact with Bantu neighbors that I think it is appropriate to classify them as Symbiotics.

Australians, however, are virilocal, patrilineal, and dominated by extended families; men hunt large game like kangaroos *and* they engage in warfare. They are joined by three other types with warfare, who also share many of the traits of the Australians: Big Game Hunters (Macrobands), Settled Fishermen, and Mounted Hunters (see figure 6.1).

A researcher who wants to answer the question as to whether hunter-gatherers have war might select a sample of hunter-gatherer societies. Depending on how many societies are in each category, the frequency of war among hunter-gatherers should be around 50 percent (four categories vs. four categories). This frequency can be varied by excluding from the sample some categories. If Australians, Settled Fishermen, and Mounted Hunters are removed from the sample, the frequency of hunter-gatherers *without* war can rise to 80 percent (four categories to one category), but if Symbiotics and Incipient Tillers are removed instead of the above three types, the frequency of hunter-gatherers *with* war can rise to nearly 70 percent (four categories to two categories). Therefore, sample selection can skew the answer to the question, and samples of hunter-gatherers show that this has occurred.

We will look at three examples of research on the question of whether hunter-gatherers have war. The first comes from political scientist Azar Gat. He focused on Simple Hunter-Gatherers, who for him are the Australians, and Complex Hunter-Gatherers, who for him are the Northwest Coast Settled Fishermen. In his words:

> To summarize the finding from our two—Australian and northwest coast—hunter-gatherer "dream laboratories," they clearly show, across a very large variety of native peoples living in their original settings, that hunter-gatherers from the very simple to the more complex, fought among themselves. Deadly conflict if not endemic war was to be expected. (2006:35)

Our second example, at the other extreme, comes from Douglas Fry who removed the Mounted Hunters and Settled Fishermen from his sample of hunter-gatherers, and he argued that the Australians feud rather then engage in war, thus leaving a sample composed of the other five types. Frequencies are predictable. "The essential finding is that all of the complex hunter-gatherers and all the equestrian hunter-gatherers make war, whereas a majority of the simple hunter-gatherers do not" (2006:104; the entire passage is italicized in Fry). According to Fry, eight of the simple hunter-gatherers have war and 13 do not; therefore, 62 percent are *without* war. If the nine complex hunter-gatherers and the five equestrians are added to the 21 simple hunter-gatherers, the percentage of hunter-gatherers *with* warfare rises to 68 percent. The above "exercise" can also be performed using virilocal vs. multilocal residence. Recall that warfare and virilocal residence go hand in hand. The results are roughly the same.

Our third and even a more extreme example comes from Elman Service's small textbook titled *The Hunters*, in which he describes the nine groups of hunter-gatherers who form his "sample" (his term) (1979:76–98). Five groups are Simple Hunter-Gatherers: the Eskimo, Indians of the Great Basin (western U. S.), Indians of Tierra del Fuego (tip of South America), Semang, and Andaman Islanders. Two are Symbiotics: Pygmies (Mbuti) and Bushmen (!Kung). The Algonkian and Athabascan Hunters of Canada are Big Game Hunters—Microbands. And, lastly, not surprisingly, the only armed combat described is for the Australians. Notice his sample includes only four of the eight types, and that only one of his types has warfare. This permits him to conclude that "warfare is exceptional at the band level of society" (1979:56). But Service does recognize that the threat of violence is always present in hunter-gatherer societies (1979).

So the answer to the question that heads this section is: "It depends." If someone claims that hunter-gatherers do not have warfare, ask the person what types he or she is examining and to give examples of specific societies. On the other hand, if the person claims that hunter-gatherers have warfare, ask the same follow-up question. Service did this nicely for us.

Does a time perspective change the frequency of hunter-gatherer warfare? I believe it does. The peoples at, say, the beginning of the Upper Paleolithic would have been Big Game Hunters and Foragers. As hunting increased, so did warfare. Then war declined as the percentage of Foragers increased due to extermination of many large animals. The settlers of Australia hunted large game, driving some animals and large birds to extinction, but the fast-leaping kangaroos survived to supply the Aborigines with sustenance. As people turned to fishing near the end of the Upper Paleolithic, villages of Settled Fishermen formed around the lakes of Europe. Later, travelers to the New World settled along the Northwest Coast and around the lakes of North America. Pockets of warfare occurred at these locations. In more recent times, after the Neolithic, horses became available to hunters who used them in the pursuit of large game, while other Big Game Hunters were driven to marginal areas where large game lived. They hunted in smaller groups and seldom warred with each other. Once horticultural and agricultural villages filled much of the land, Foragers were able to obtain seeds and the like, which turned them into Symbiotics and later Incipient Tillers. Warfare, had it occurred, would have destroyed their relationship with the villagers and cut off their supply of produce and perhaps tools made from iron.

PRIMARY STATE WARFARE

Some researchers believe that warfare arose first in early farming villages in the Middle East and elsewhere. Competition for fertile fields and for the produce from these fields started humankind on its long climb to global warfare. World War II, if not World War I, was global warfare. Many books on the development of warfare start with hunter-gatherers, who are said either not to engage in warfare or to participate in ritual warfare. As we have just seen, their authors are incorrect. These researchers, from many different disciplines, argue that the first combat arose between early farming villages and that it was ritual warfare. Thus, warfare, according to these researchers, first appeared in the early Neolithic (e.g., Ferguson 2006; Haas 2001).

They argue that as the villages became larger the warfare grew more serious; fatalities rose as warriors became intent on killing each other. If one village conquered another, this larger village could proceed to incorporate smaller villages into its growing political structure. After chiefdoms formed, and one chiefdom conquered another, there were three rather than two hierarchical levels to the political structures. Some archaeologists call this a state, others wait until one of these three-level polities has conquered another three-level polity. A state, thus by this definition, has four hierarchal levels. There is no consensus among archaeologists as to whether a state, by definition, is a three- or four-level polity. This, however, is not important. What is important is that the excavators ascertain how many political levels there were.

The conjectural sequence that I have just described is called the conquest theory of the state. It has deep roots that go back before the 19th century, but it was in the latter half of the 19th century, as European nations began to conquer native people and sometimes each other (e.g., Franco-Prussian War, [1870]), that the theory was fully developed. The conquest theory of the state is described in many thick archaeology and anthropology textbooks, but it is not accepted by all scholars.

The reason this theory is controversial is that some scholars, including myself, think it has a major logical flaw. Archaeological data from excavations in the areas where primary or pristine states arose do not support the theory. The flaw lies in the argument that "one village would conquer another." Villages do not have a political structure that permits conquest and incorporation. The polity has a headman and/or a council of elders with only limited power. Furthermore, the political leaders do not harbor an ideology of conquest and incorporation. Unless they have seen it elsewhere, it would be unlikely to occur to them. Remember why uncentralized political communities go to war—after land, plunder, and women. There is no government to do the incorporation. There is no political way the two villages could be joined together. The results of archaeological excavations in regions where pristine states arose suggest how the earliest forms of centralized government arose. Before I tell the story of how the first states arose, I need to describe the origin of agriculture. And here I will be telling a story that I believe has not been told before.

Origin of Agriculture

Agriculture means the growing of plants and crops that we and other animals can eat. It is something deliberately done. Merely picking berries, seeds, and the like is gathering. For gathering to become agriculture several conditions must occur: the seeds or whatever the crops grow from, must be planted; animals and birds need to be kept from eating and destroying the crops. This is a grade school social studies explanation for the origin of agriculture. A professional description does not differ much.

One thing that the descriptions do not mention, which I think is crucially important, is that warfare must be absent from the region where plants are domesticated. If warfare occurs, domestication cannot take place. Raiders will kill and drive off the settled gatherers who are tending the plants, steal the seeds intended for planting and any produce already in storage if there are captives who can carry containers, and destroy the fields and buildings. Frequently buildings and granaries are set on fire.

Now we take a closer look at the origin of agriculture. A selectionist explanation is preferred by many archaeologists and biologists. "Initial domestication" occurred as the result of seed dispersal, perhaps deliberate, and the protection of wild plants. Later, people selected consciously for size, bitterness, fleshiness, and oiliness. As domestication got underway, people selected unconsciously for "invisible" features, such as dispersal mechanisms, germination inhibition, and reproductive biology. "Specialized domestication" occurred when people began to selectively destroy unwanted plants and to protect those they wanted because of desirable features, such as those mentioned above. In the third stage, practices such as weeding out undesired plants, irrigation, and fertilization created further plant evolution (Henry 1989). The rate at which wild plants could be domesticated varied greatly. A few hundred years for wheat in the Middle East and about 2,000 years for maize or corn in Mesoamerica. This brief description of how plant domestication probably occurred shows how vulnerable the fields and gardens were to enemy attack (Otterbein 2004).

In addition to the absence of warfare and the presence of wild seeds that can be domesticated, there must be fertile soil and ample water to nurture the start of the germination process and to keep the plants growing. But most important of all are the Foragers who settle in one location and become Incipient Tillers. About 15,000 years ago foragers began to develop facilities for storing seeds, made grinding stones, and developed the technology for processing seeds. By 10,000 BCE sedentary communities existed along the Fertile Crescent, which arched from the Levant to southern Mesopotamia. These settlements did not engage in warfare.

On the upper Euphrates River in about 9500 BCE the village of Abu Hureyra was established by settled gatherers. A thousand years later it was a large farming community. The people replaced their circular dwellings with rectangular houses. By 5000 BCE the settlement had grown to nearly 6,000 residents, when aridity forced its abandonment. What is significant about Abu Hureyra is that there is no evidence for warfare for the entire 4,500-year period. This conclusion is based on an extensive archaeological site report (Moore, Hillman, and Legge 2000). This is both my conclusion and the conclusion of the chief investigator, Andrew Moore, with whom I have discussed Abu Hureyra. In southern Mesopotamia neither warfare nor aridity prevented the development of the first state, Uruk/Sumer in 3500 BCE. Domesticated plants had spread there from the Fertile Crescent by 5700 BCE.

Warfare along the Nile River before 10,000 BCE, going back as far as 20,000 years ago at Jebel Sahaba, prevented the process of domestication from starting in that region. Not only can warfare prevent the development of plant domestication but it can also prevent the development of the state. I argued above that primary states do not arise though conquest. Indeed, I could have also argued in that section that warfare between warring villages not only did not lead to conquest and to statehood, it also prevented the development of the state. Warring villages can destroy each other—consider the Dani and Yanomamö and what is often the fate of their villages. And even if they do not destroy each other, warfare prevents centralization from developing because the headman and the men of the villages are regularly killing each other. As shown at the end of chapter 2, prisoners are usually not taken, and if they are, they will soon be killed (sacrificed in some cases) or sold as slaves. Villages, due to population loss, do not grow in size. The destruction of fields and other resources, such as animals, water holes, and granaries, prevents village growth. In Anatolia or Turkey, warfare developed after 7000 BCE. Control of trade routes and the sources of obsidian seem responsible. Obsidian is choice weapon material for projectile points and daggers. Warfare prevented state development. Just as war prevents the domestication of plants, war prevents the development of the state.

Four Primary States

The earliest states in several world areas have long been recognized. In addition to being the first states in their region they are far apart geographically. Geographic separation means each developed without influence from the others. I have seen lists in the past with six states. Two of these I have not used; four pristine states that I have used in my analysis are: Zapotec of Central America; Chavin/Moche of North Coast Peru; Uruk/Sumer of Mesopotamia; and Hsia/Shang of northern China (see figure 6.2).

How *did* the first, or primary, states arise? We can infer from archaeological site findings how each of four states arose. When needed, we use "bridging arguments," techniques for inferring what we are looking at. For example, an archaeologist has uncovered a young male laid out with breastplate, shield, sword, and spear. A projectile point lies between two ribs. In "implicit ethnographic analogy" the researcher infers that the man must have been a warrior, probably an elite warrior since an elaborate, unique helmet had been placed beside him. He had been shot by an arrow or atlatl dart. The size of the point, computed by multiplying width times thickness divided by 2, can be compared to existing scales, which give the size range for arrow and dart tips. Comparing the point to these scales permits the archaeologist to conclude which weapon killed our man (Shea 2006).

The procedure I followed was to gather from archaeological studies focusing on each area information that described each stage or phase that the society had gone through on its way to statehood. These phases are typically given regional names by their excavators. Information was sorted into four categories: Social and Political Organization, Internal Conflict, War, and Military Organization. These topics have been the subject of the early chapters of this book. For each society, the phases were used to form steps of a "ladder," and the categories formed four columns that cut across the phases of state development (see figure 6.3 on p. 81).

To compare the cultures it was necessary to construct a ladder of steps, with each step being a developmental stage. The phases for each society were fitted to the evolutionary ladder, which overlapped two typologies that have an evolutionary sequence built to them. Robert Carneiro's typology of chiefdoms (1981) was combined with Henri Claessen and Peter Skalnik's typology of early states (1978). For Claessen and Skalnik an early state is a newly formed state, no more than approximately 200 years old. The combined typology has the following appearance:

Stage V   Transitional Early State

Stage IV   Typical Early State

Stage III   Maximal Chiefdom = Inchoate Early State

Stage II   Typical Chiefdom

Stage I   Minimal Chiefdom

This typology fits between Agricultural Villages and the Mature State. Seven steps in all. In chapter 3 the Despotic State and the

Stages	Zapotec	Chavin/Moche	Uruk	Hsia/Shang
Mature State	**Monte Alban III** Elite warriors	**Chimu** Elite warriors Sacrifice	**Akkadian** Infantry and archers Codified law	**Chou** Massed infantry and cavalry
Transitional Early State	**Monte Alban II** Elite warriors Forced relocation	**Middle Horizon** Elite warriors Tensions—cities abandoned	**Early Dynastic III** Massed infantry and chariots	**Shang** Elite warriors in chariots
Typical Early State	**Monte Alban I** Elite warriors Forced relocation	**Moche** Elite warriors Locals tortured Sacrifice	**Early Dynastic I & II** Chariots and infantry Slaves	**Hsia** Elite warriors
Inchoate Early State	**Rosario** War (late) Forced labor	**Chavin** No war Cannibalism?	**Uruk** Elite warriors War (late) Coercion	**Longshan** No war? Internal violence
Typical Chiefdom	**Guadalupe** No war Minimal conflict	**Initial Period** No war Minimal conflict	**Late Ubaid** Peaceful trade Uprising	**Miao-ti-Kan II** No war? Internal violence?
Minimal Chiefdom	**San Jose** No war Minimal conflict	**Late Pre-Ceramic** No war Minimal conflict	**Early Ubaid** No war Minimal conflict	**Yangshao** No war Minimal conflict
Agricultural Villages	**Tierras Largas** No war Minimal conflict	— No war Minimal conflict	**Eridu** No war Minimal conflict	**Peiligang** No war Minimal conflict

Figure 6.2 Development of Four Pristine States

Mature State are described; Early States are Despotic States, particularly at Stages III and IV. To my surprise, for each primary state this typology fit over the archaeologists' named phases. This information is shown in figure 6.2. Detailed descriptions of each pristine state are in my book *How War Began*, along with maps and the figures that show the content of each named phase (2004:121–176). For each of the four states, the phases that were at the same stage were similar in content. This allowed me to derive generalizations of how the first states, taken as a group, developed from being settled gatherers to becoming large culturally homogeneous states. The content of each stage for each category is listed in figure 6.3.

The generalizations describe the second path that warfare has taken. Recall from our discussion of types of hunter-gatherers, that in the areas where large animals became extinct, the Big Game Hunters became Foragers and then Settled Gatherers. Warfare no longer occurred. Wild plants became domesticated and Agricultural Villages arose, surrounded by fields of grain and other crops. Wheat in Mesopotamia, millet in northern China, corn in Mesoamerica, and the potato in Peru. Warfare was absent. There were headmen and kinship groups but no military organizations, and internal conflict was minimal.

At the Minimal Chiefdom stage the villages grew larger, the political leader became a chief, wealth differences arose, and conflict began between the leader and want-to-be leaders. Nevertheless, there was still no war or military organization.

At the Typical Chiefdom stage the village became so large it fissioned and a two-level polity formed. Wealth differences became so extreme that social classes formed and violent clashes between leaders occurred, which included assassinations. There was still no war, but Fraternal Interest Groups arose whose members carried knives, spears, and clubs and threatened others. Fortified residences were built. A Type A polity emerged.

At the Maximal Chiefdom = Inchoate Early State stage, life gets tough. Neolithic bliss has ended. The chief becomes a despot, the leader of a three-level polity, whose own village becomes a "capital." Several states in the same culture may arise. A pronounced class system exists. The chief's henchmen seize and execute rivals, and also members of the lower class, just to terrorize the population. Human sacrifice occurs. The military is composed of elite warriors. Villages are fortified. Internal war is born. Wars of consolidation take place between close neighbors. Changes can be observed within this stage. Initially there is an internal struggle that is very violent. When one "thug" has become chief he almost immediately attacks neighboring polities. Any "enemy" leaders captured will be executed or sacrificed. Near the end of this stage a four-level polity is formed and the chief becomes a dictator. This was a Despotic State with Elite Warriors.

Stages	Social and Political Organization	Internal Conflict	War	Military Organization
Mature State	• Monarch or representative leader • Empire with culturally different peoples/members of polity	• Bureaucracy replaces despotism • Judicial system with laws	• Alliance wars	• Professional commander • Light infantry provided by different peoples
V. Transitional Early State	• Monarchy with royalty, upper class, commoners, and slaves. • Four levels to polity: capital, cities, towns, and villages	• Consolidation of same-language polities	• External war	• Distinct military units: massed infantry, troops with projectile weapons, siege train • In Old World, chariot corps
IV. Typical Early State	• King heads military aristocracy • Four levels to polity: capital, large towns, small towns, and villages • Slavery	• Lower class taxed—agricultural surplus and labor • Men conscripted • War captives enslaved • Human sacrifice	• Internal war	• Professional army: aristocratic officers; conscripted commoners form massed infantry with shields • Fortified cities and siege operations
III. Maximal Chiefdom = Inchoate Early State	• Chief is despot • Chief's village becomes "capital" • Three levels to polity: capital, towns, and villages • Upper class dominates lower class	• Execution of rivals, members of lower class • Human sacrifice	• Wars of consolidation—close neighbors	• "Army" composed of chief, followers, and upper class males • Shields and helmets • Village fortifications
II. Typical Chiefdom	• Chief from upper class, supported by kin • Two levels to polity • Social classes	• Violent conflict between leaders • Assassinations	• No war	• Fraternal Interest Groups • Spears, clubs, and knives • Fortified residences
I. Minimal Chiefdom	• Chief • Large villages • Wealth differences	• Conflict between leaders begins	• No war	• None
Agricultural Villages	• Headman • Kinship groups with leaders • No wealth differences	• Minimal conflict	• No war	• None

Figure 6.3 Stages in the Chiefdom–State Continuum

At the Typical Early State stage, the political leader, now a king, heads a military aristocracy. The lower class is taxed and young men are conscripted into the military. This is a professional army run by aristocratic officers. Commoners form massed infantry with shields and spears or pikes. Cities are fortified and siege operations are perfected. The basic pattern of warfare is lines and siege operations. This is still a Despotic State with an army led by elite warriors controlling conscripts. They are viewed as rabble—the equivalent of "cannon fodder."

At the Transitional Early State stage the four-level polity develops a four-class system with a monarch and royalty at the top; below is an upper class, which sits above commoners and slaves. All polities with the same language have become consolidated into one kingdom. Other kingdoms with different cultures and languages are attacked, or they attack the polity that is our focus. External war in both of its manifestations arises. The military organization becomes perfected. Integrated Tactical Systems arise, often with heavy massed infantry, light infantry with projectile weapons, and, in the Old World, chariot corps and cavalry.

At the Mature State stage an empire is likely to have formed, composed of culturally different conquered peoples. A bureaucracy replaces despotism and a judicial system with laws and courts arises. People may be "citizens" with civil rights. The military may have within it light infantry provided by culturally different peoples. But the Mature State still has a conscript-based army. The polity may enter into alliances for the purpose of waging war against other states, particularly early or despotic states.

The path just described pertains to primary or pristine states. After I identified this path and recognized how the stages flowed into each other, I attempted to find an explanation for the transitions. A number of researchers including myself have developed internal conflict theories, which are reviewed in *How War Began* (2004:101–105). My internal conflict theory rests on Fraternal Interest Group theory and pertains to the Chiefdom stages. A second theory is needed for the Early State stages, for which political legitimacy theory, from sociology, serves our purpose. Ironically, punishment is used to force compliance, which in turn legitimizes the polity. Sociologist Steven Spitzer states:

> Pristine states, precisely because they lack legitimacy, must develop and impose harsh, crude, and highly visible forms of repressive sanctions; developed states, having successfully "re-invented" consensus, can achieve social regulation through a combination of civil law and relatively mild forms of "calculated" repression. (1979:210)

Now you know "how war began" along two separate paths.

Chapter Seven

The Evolution of War

When culturally different peoples engage in warfare, a variant of natural selection theory can be used to explain efficiency in weapons and tactics, the military success of some polities, and the survivability of polities. This variant of Darwin's theory is known as "cultural group selection." The groups can range from hunting-gathering bands to modern states. They compete with each other and the more successful group will grow in population, create more material culture, and probably come to occupy more territory. Sometimes parts of bodies of water or marshes are filled in to expand an island or to create fields or residential areas where ponds and lakes are located, for example the peninsula of Charleston, South Carolina (heartland of Confederate rebels), or the Valley of Mexico (capital of Aztec empire). When the competition between polities involves armed combat, warfare between polities may become a life and death struggle. For the loser, this can involve the destruction of the polity and the killing and enslavement of survivors. A "fortunate" few may escape to other polities, becoming refugees.

Fitness in the natural selection of Homo species pertains to the ability of the larger, stronger, healthier, more intelligent hominids to survive long enough to produce offspring. In cultural group selection the larger, stronger, "healthier," and most capably managed polity is likely to remain viable longer than competitors. Competition between groups, we have seen, contributes to cooperation within the group and to ethnocentrism. Ethnocentrism facilitates socialization for war. But it is the military ability of the polity, in terms of the quality of organization, tactics, and weapons, which usually determines the outcome of a battle or campaign and the fate of the polity. In warfare, the polity with the more efficient military practices, as measured by a Military Sophistication Scale, would expand territorially (Otterbein 1970). The greater the number of efficient practices, the higher the scale score. Similar practices were compared. For example, professional military organizations are deemed more efficient than nonprofessional military organizations; it is better to have both projectile and shock weapons than just one type of weapon; and integrated tactical systems trump simple lines. Many more traits were used to construct the scale (C. Otterbein 1970).

WARFARE SYSTEMS

I began my study of warfare in the early 1960s by constructing a framework that would apply to both case studies and cross-cultural studies. It was a theoretical approach that focused on intergroup relations. Rather than select a single entity for study, such as a kinship group (like a Fraternal Interest Group) or a political unit (like a political community), I selected two or more entities and the relationship between or among them for study. This approach is drawn more from sociology, political science, and social psychology, than from anthropology. The Robbers Cave Experiment was one of my sources. Not only was I interested in examining what happened to two warring polities, I was interested in what transpired *within* them as they battled. My framework, theoretical approach, or model also has a time dimension.

The warfare that occurs over a period of time between two or more rival entities (political communities or alliances) constitutes a "warfare system." The framework described above is useful in analyzing warfare systems. The approach was used with success in a cross-cultural study of feuding (K. Otterbein and C. Otterbein 1965). Here it was discovered that uncentralized political systems with Fraternal Interest Groups had both feuding and warfare while centralized political systems that engaged in warfare suppressed feuding. The distinction between internal and external war had yet to be made. The approach was also used in case studies. Their subjects explain: Huron versus Iroquois; Nuer versus Dinka; and Zulu versus other Nguni polities (Otterbein 2004; their original publication dates stretch over many years). In each case study the militarily more efficient polity crushed its opponent or territorially expanded. Iroquois burned Huron villages, Zulu incorporated other Nguni tribes into its kingdom, and the Nuer expanded territorially as they raided for Dinka cattle and women.

Even cultures without war may be part of a warfare system. For example, the Semai, the Incipient Tillers of Malaya who did not engage in warfare, have been attacked so frequently that they have become a refugee group or enclaved people. Dentan speaks of a "geographic refuge" inhabited by peoples who adapt to slave raids by their ability to disperse and regroup. The Semai became nonviolent and developed values of peaceability (1992). Another reason a culture may be without war but part of a warfare system is that it is in a symbiotic relationship with a larger, usually politically complex culture. Two hunter-gatherer bands that became Symbiotics, the Mbuti and the !Kung, are examples. A third reason is that it has been conquered by a state and become a Dependent Native People (Otterbein 1977). Such encapsulated cultures often rise up and attack their conquerors. Today such an uprising is referred to as an insurgency and the people are called "insurgents." Efforts to suppress an insurgency can range from genocide to social reforms. This happened in Ancient Times, during the Colonial Period, and I am sure it is happening somewhere as you are reading this passage. Totally isolated cultures would not be part of a warfare system unless the peoples on, say, an isolated island, like Bellona, had formed polities. Then the entire island is a warfare system.

The notion of a warfare system is useful to archaeologists. Rather than focus on a single site, the focus is on a region (if resources permit). The first task of the archaeologist is to identify the polities within a region and to attempt to ascertain their relationship to each other. Using a diachronic approach, the archaeologist may be able to determine settlement locations and movement over time, which settlements have expanded and which have contracted in size, which have fissioned, and which have coalesced. Walls surrounding settlements and settlements located on a hill or in a place that is difficult to find and approach suggest the presence of warfare. The report contained in *Zapotec Civilization* encompasses the entire Valley of Oaxaca and follows the fortunes of the warring peoples for a millennium (Marcus and Flannery 1996). Hidden homesteads suggest fear of attacks connected to feuds. Fortified houses, compounds, and neighborhoods indicate fear of crime, attacks, or an uprising against the upper or ruling class; a burned settlement may indicate warfare, while burned houses indicate an uprising (Haas 2001; Otterbein 2006b). If future archaeologists were conducting a survey of several excavations in Higiland, they would probably conclude that the stone huts, stone compound walls, and location on the sides of the Mandara Mountains suggest local crime, feuding, and war. And they would be correct.

Alliance Systems

Alliance systems are warfare systems. Polities generally enter into alliances for the purpose of security or defense; it is a way for a polity to protect itself in a dangerous world. Leaders hope that, if an enemy attacks their polity, an ally or allies will come to its aid militarily. If a polity has neighbors with whom it cooperates in trade, visiting, exchange of men and women, and/or ceremonies, the likelihood of armed combat between them is low. These cooperating polities may be allied or potential allies. This is true if political leaders are on cordial terms and if the sides do not have negative images of each other; in other words, ethnocentrism is at a minimum. Buffer zones need not arise and fortifications need not be erected along boundaries. My home in western New York state was in a war zone in 1812; there are many battlefields and forts along the Canadian–United States border. In the 21st century they are tourist destinations. Those of us who live in either Ontario or New York State do not have negative images of each other.

Alliances are often formed in order to prepare for a joint military venture. The allies may form a huge army combining commands, or they may attack from separate directions. This was the dilemma Frederick the Great faced in the 18th century (Ritter 1968). He was surrounded by three allies: France, Austria, and Russia. He was forced to develop an aggressive military strategy. He would attack the armies of first one, then another nation, usually winning. This strategy has come to be known as fighting on *interior* lines. The Allies fought on *external* lines. Frederick's only ally was England, and England was involved in a major war in Scotland, which culminated in the Battle of Culloden in 1746.

But it is not only in recent centuries that alliances have become important. They were also important in uncentralized political systems, probably through millennia. Most of chapter 1 was devoted to describing the different alliance systems of the Dani and Yanomamö. I chose them for their renown and their movies. I could have chosen them just as easily for their warfare systems. They are similar in many ways, yet strikingly different. The Dani I have described as having a "balanced" alliance system and the Yanomamö an "unbalanced" alliance system. Diagram their systems and you will see why I chose these terms. (If you want to see my diagrams see *The Evolution of War* [1985a or 1989b:xx–xxi]). Hint: one system looks like an equilateral triangle and the other an equilateral triangle that is missing one side.

Warfare systems can be classified as closed or open, depending on whether armed combat is between culturally similar polities or culturally different polities. While these terms are part of the language of general systems theory, I am using them in a different sense. Anthropologist Robert Harrison, following Andrew Vayda, thinks of the Maring of New Guinea as having an open system (1973), while, in the sense I use the terms, it is a closed system. The warring polities, being only Maring, are culturally similar. Their primary ethnographers, Roy Rappaport (1968) and Andrew Vayda (1976) have presented different analyses of Maring warfare.

Closed Systems—Internal War Only

A *closed warfare system* is not only bounded and isolated; the polities within are culturally similar. It is a system in which only internal war occurs. The island of Bellona is an example. Culturally similar polities in an isolated region may engage in warfare for a long period of time without making changes to their weapons and tactics. Their battles resemble duels. The four characteristics of duels that I identified seem to pertain: (1) lethal weapons are the same on both sides; (2) rules are followed and battles are arranged; (3) the same social class is represented on both sides; (4) motives are honor and revenge (Otterbein 2001). Greek phalanx warfare, which was practiced for hundreds of years, underwent little or no change, and, like a duel, was highly rule-bound.

A closed system can arise when one culture occupies a body of land surrounded on most or all sides by water. The Peloponnesus is an example. For over 1,000 years, perhaps longer, Greek culture more or less uniformly covered the peninsula; this period extended from before the end of the Bronze Age in 1200 BCE until the Persian Wars in the 5th century BCE. The exploits of the hoplite phalanx are known to any war lover or, more politely, military buff. Sometime after the period of the Iliad (see chapter 3), its origins still seemingly unknown, a heavy infantry developed in the Greek city-states. A hoplite carried on his left arm a round shield more than three feet across, in his right hand a seven- to eight-foot thrusting spear, and a short sword in a scabbard; he wore a corselet, a metal helmet, and greaves to protect his legs. In battle the spear was carried in an over arm position. The soldiers provided their own arms and weapons. This was a middle-class army. Battles between armies took place on level ground between the city-states. "The two opposing phalanxes met each other with clash of shield on shield and blow of spear against spear" (Adcock 1962:4). The phalanxes charged each other once, twice, three times until one phalanx broke and took to flight. Then the battle was over. The victors took possession of the field of combat and any weapons left on the field or on the bodies of the dead, and the losers fled to their fortified cities (Ducrey 1985).

For our purposes we are looking at an area where military change occurred slowly over 500 years (about 800 BCE to about 300 BCE). Every polity was nearly the same. The rules and customs were known by each side. Lendon, in *Soldiers and Ghosts* (2005), argues that the heroic values the Greeks developed in the Dark Age (about 1000 BCE to about 750 BCE), are exemplified in the *Iliad*, written perhaps just before the Olympic Games began in 776 BCE. The heroic values persisted throughout the period of hoplite warfare. Although individualism was suppressed in phalanx warfare, it still served as an ideal to guide men. Only near the end of the hoplite period, and well after the Persian Wars, did sophisticated tactics develop. A general named Epaminondes placed his troops in depth on the left wing so that they would overwhelm the right wing of his opponent's line. At the Battle of Leuctra, in 371 BCE, Thebes defeated Sparta. Epaminondes was killed at his second great victory, the Battle of Mantinea in 362 BCE.

In a closed system, military change is likely to come slowly. No polity is likely to conquer and expand. Only the arrival of outside influences and forces is likely to change the system. The culturally different Persian army invaded Greece and the Persian Wars began.

> But even when strategy and tactics are, as it were, conventional in wars between Greeks, it may be otherwise when new enemies and new methods of warfare force themselves on the attention of the city-states. And this the Persians did in the two enterprises that led to the battle of Marathon [490 BCE] and then eleven years later, in the battle of Plataea [479 BCE]. The Persians had conquered from Iran to the Levant by the skill of their cavalry and archers. (Adcock 1962:11)

Striking at the critical moment, the hoplite phalanx won the day for the Greeks at both battles. Thus, hoplite warfare continued for over 100 years more into the Peloponnesian Wars.

This lack of military change in a closed system seems to characterize the warfare of many nonliterate people for a long time prior to the expansion of state-level societies. The Dani and the Yanomamö are prime examples. So are the Maring. Their warfare systems seem timeless in that they appear to have no beginning, and they seem as if they could go on forever. They do not, however, because Europeans enter their regions. In the case of the Dani, they are pacified by the Dutch and the Indonesians; in the case of the Yanomamö, explorers and missionaries, and some would say anthropologists, created more warfare.

Open Systems—External War

A warfare system in which the armed combat is between culturally different peoples is an ***open warfare system.*** Cultural group selection theory explains improvements in weapons and tactics, as well as which culture will be militarily successful, only for warring polities that are culturally different. The theory pertains to external war, not internal war. When two culturally different polities seek to defeat each other, the desire to improve or invent new weapons and tactics arises as each polity attempts to achieve military supremacy. Changes in weapons and tactics can occur rapidly. Since changes are likely to be in the direction of greater effectiveness, the military organizations that have made the changes are likely to be victorious on the battlefield. Cultures with high military efficiency have expanded territorially (Otterbein 1970).

We now return to the evolution of Greek warfare. The attacking Persian army was composed of entirely different military units, which were known to the Greeks but not employed on the Greek battlefields. They were cavalry and archers. Chariots had largely disappeared from the battlefield at the end of the Bronze Age. Robert Drews argued that light-armed troops attacking the chariots with throwing spears drove them from the battlefields and brought the Bronze Age to an end by

1200 BCE (1993). More recently Drews has argued that by the 7th century BCE cavalry made its appearance, a date much later than many scholars have contended (2004).

Greek warfare was transformed. Cavalry, archers, and light infantry known as peltasts joined the military. Peltasts carried small round shields and throwing spears. They were often mercenaries from other ethnic groups under the command of professional officers. Peltasts were used in rough terrain and served as skirmishers in front of the hoplite line. Cavalry also served on the flanks, and archers could be used to provide firepower from behind the line. Combined arms tactics developed (Ducrey 1985).

Philip of Macedonia (381–336 BCE), the land to the north of the Peloponnesus, took the new tactics further. The soldiers in the phalanx were equipped with narrow rectangular shields and long pikes held at waist level with two hands. They did not charge; with a steady advance, they could crush their enemy. Philip's son, Alexander (356–323 BCE), perfected the combined arms tactics by using the phalanx and light infantry to open a gap in the enemy's line. The cavalry galloped through the gap and headed toward the enemy general/monarch at his command center. Victory could come quickly. Alexander's numerous battlefield successes rested on the finest logistic system the ancient world had seen (Engles 1978). Some military historians, like Arther Ferrill (1985), contend that Alexander created the finest military organization the world has ever seen, capable of defeating the Duke of Wellington at Waterloo. Recall that Napoleon was unable to do this.

The following example, provided by Tom Wintringham, also helps to interpret the warfare of culturally different peoples engaged in armed combat over extended periods of time. The approach fits well into the Open Warfare System framework described above. The development of European military methods is characterized by alternating periods of armored and unarmored warfare. The shifts from one period to another were caused by changes in the striking power of weapons, the protection of armor, or the mobility of the armies. When weapons became so powerful that they could penetrate armor, protection was abandoned and mobility became an important element. Until armament improved or tactics based on high mobility were devised, the side with the more effective weapons would be victorious. Eventually the pendulum would swing back when armor once again could efficiently stop the firepower of weapons (Wintringham 1943).

I used Wintringham's method to analyze Iroquois warfare. The Iroquois went from a period of armored warriors to unarmored warriors, after firearms were introduced by the English to the Iroquois. They fought the Hurons and the Algonquians throughout the early half of the 17th century, achieving final victory in 1649 (Otterbein 1964b, 1979b). I also used a similar approach in a study of Zulu warfare

(1964a). The Zulu developed a new military organization (regiments from their system of age-grades), new tactics (the chest-and-horn formation), and a short stabbing spear (to replace several throwing spears). In Zululand great mobility was achieved by reducing the size of the shield and the number of spears carried as well as devising regiments that could act as mobile units. Warriors now ran.

In Mesopotamia, about 3000 BCE, four-wheeled carts were drawn by equids (perhaps a sterile cross between a donkey and a wild onager); the driver and an elite warrior armed with javelins and a socket ax drove into the enemy forces. Infantry with spears, helmets, and studded capes followed. In a later period, the ruler or commander rode alone in a two-wheeled chariot, still drawn by equids, with javelins in a quiver, and he carried a bronze sickle-sword. On the field of battle, he led on foot a column six files deep. Each soldier carried a rectangular shield and a long spear. When the warriors are placed in a phalanx, their spears extend forward of the front wall. By approximately 2000 BCE the Mesopotamians were obtaining from pastoral peoples to the north domesticated horses that could be harnessed to their chariots (Otterbein 2004).

In north China, horse-drawn chariots were introduced to the Shang from the west in about 1200 BCE. The chariot was fully developed when it arrived. A large chariot, it carried three soldiers: a driver, a striker with a halbert (small ax blade on a long handle), and an archer. I have identified three phases to Chinese chariot warfare, which lasted until about 300 BCE, when massed infantry armies, based on conscription, and cavalry came to dominate the battlefield. In the first phase, elite warriors and rulers were transported to battlefields but continued to fight on foot. In the second phase, combat took place on a flat, obstacle-free, agreed-upon location, between chariot crews drawn from the nobility. In the third phase, chariot squadrons performed specialized military tasks, such as transportation of officers, reconnaissance, special maneuvers, pursuit of defeated enemy, and rapid retreat by monarch and nobility in case of defeat (Otterbein 2004).

After describing chariots in many Old Word regions, historian Arthur Cotterall compares the war chariot to the introduction of the tank in World War I and concludes they are totally different weapons. Unlike chariots, tanks are closed up, provide poor visibility, and make mind-numbing noise (2005). A better comparison would be to an open air, lightly armored and minimally armed motorized vehicle, such as the U.S. Army's High Mobility Multipurpose Wheeled Vehicle (M998 Truck—HMMWV). Means of transport, often introduced by another culture, can transform warfare.

Our final case of military change that arises from the contact of cultures is Frank Secoy's study described in *Changing Military Patterns on the Great Plains* (1953). Secoy shows how the **diffusion** of

horses northward from the Southwest influenced Indian warfare practices in western North America. Warriors became mounted while con-
tinuing to use their traditional weapons: bows and arrows and spears.
Next, he shows how the diffusion of firearms westward from the Atlan-
tic seaboard influenced Indian warfare east of the Mississippi River.
Fighting from ambush, with the warriors often deployed in long lines,
was the most efficient way of using the slow-loading muskets. The
"post-horse—pre-gun pattern" and the "post-gun—pre-horse pattern,"
as Secoy calls them, met in the Great Plains, and a new mode of warfare developed, called the "horse and gun pattern." Warriors equipped
with guns went to war mounted on horseback. Secoy has a series of
maps that show how far the horses had traveled from the southwest
northward in the west and how far the guns had spread from east to
northwest at different time periods.

Secoy's study also well illustrates tribal zone theory, developed by
anthropologist Brian Ferguson. Numerous case studies are presented in
War in the Tribal Zone (Ferguson and Whitehead 1992). The "tribal
zone" is an "area continuously affected by the proximity of a state, but
not under state administration." The main effect of state intrusion is to
increase "armed collective violence" (1992:3). In this situation, three
categories of war can occur (1992:19–25). First, wars of resistance and
rebellion. This occurred on the Great Plains after the era covered by
Secoy's study. The most famous example is the Battle of the Little Big
Horn. Second is "ethnic soldiering," where native combatants are under
the control or influence of states. Third is "indigenous warfare." The
Plains Indians fit in this category during the period described by Secoy.
Wars were often fought over obtaining western goods, particularly trade
in weapons. For the western Indians this included horses. Population
displacement as a result of war, in turn, leads to even more conflict.

SECONDARY STATE WARFARE

Internal conflict and political legitimacy theories, discussed in
chapter 6, seem to explain the rise of primary states. But how do all
other states arise? Do the same theories explain the origin of secondary
states? Do tribes become states? No, because while a tribe can kill the
members of another tribe that is either culturally similar or culturally
different and burn its villages, a tribe does not have the governmental
organization that would permit incorporation of the losing tribe into
the victorious tribe. Primary states, however, can conquer and incorporate other peoples regardless of the political organization of the
attacked. The rise of secondary states can be viewed within the open
system framework. The next section will briefly describe several ways

this may occur, while the following section will present two case studies that show one way this has occurred.

How Secondary States Form

The idea of statehood is crucial to state formation. With primary states it *emerges* from within a culturally similar group geographically separate from other groups; with secondary states, the idea is acquired by one or more persons who, under the right circumstances, are able to create a state. Henri Claessen and Peter Skalnik in their comparative study found that only two factors were common to all the Early States in their sample: the ideology of statehood and a surplus. Basic to the creation of a state is the idea of statehood and the notion of sovereignty: the ruler believes he has a right to rule and that his subjects owe him allegiance. Newly formed states, either primary or secondary, are known as Early States.

More crucial to state formation, it appears, is the *ability* to produce a surplus. The coercive ability of the Inchoate Early State forces lower-class members to produce a surplus, whether it be food or weapons or other goods, for the use of the upper classes and their government. These demands on the producers creates a tributary economy. A contributor to the Claessen-Skalnik study was Donald Kurtz, who analyzed the Aztec in terms of political legitimacy theory (1978). More recently, Kurtz has described the above process as the *vertical entrenchment* of the state (2001:174–188).

I contend that a surplus did not create the state, but it was necessary for the continued growth of the state. I have found a third factor common to all the early states: an efficient military organization; a surplus may be needed to achieve the necessary military excellence (1985a:xxiii–xxiv). While there are three factors necessary for a state to exist, they form a hierarchy of importance: first, the idea of statehood; second, a surplus; and third, a strong military organization to maintain the political independence of the state.

The idea of statehood can spread or diffuse from a primary state to a developed uncentralized political system and push it along the road to statehood. This appears to have happened to the Mtetwa, a Nguni clan. Their leader, Dingiswayo, had become acquainted with European explorers and Portuguese slave traders along the east coast of Africa. Specifically, some sources say that he served as a guide for a military surgeon, Robert Cowan, who was traveling overland to the Portuguese settlement at Delagoa Bay (Morris 1965). Shaka, a Zulu, served in Dingiswayo's army and may have gotten the idea of statehood also from the Portuguese, but documentation is lacking. A well-documented case, however, comes from Tonga. A shipwrecked sailor, Will Mariner, a young Englishman, with his knowledge of European military technology, helped one of the Tonga chiefs conquer other chiefs and their territories. On his return to England he wrote a memoir. Content is summarized in a chapter in a book on the South Seas titled "Will Mariner, the Boy Chief of Tonga" (Michener and Day 1958). It is an exciting tale. The old and new styles of warfare with cannons are described, as well as political assassinations. The Tonga chief learned new ideas from Will Mariner about conquering others.

States can also arise through the fissioning of primary states. A breakaway province can become a new state. A subordinate of a king, of course, knows how a state-level government is set up. The want-to-be ruler needs to rapidly create a surplus and an excellent military organization to maintain independence. Once he has done this, he is capable of creating a new Early State. Such a state will be despotic, and the despotism may persist for several generations of rulers, perhaps for as many as 200 years.

A third, less common, way for a state to arise is through fusion, meaning that polities coalesce or come together over a long period of time. Warfare plays a role, but the polities do not conquer each other. Each case I know is different. Most familiar to us is the United States. Thirteen colonies formed a confederacy after rebelling against England. Later other colonies (= states) joined the union that followed confederacy. Finally territory was either purchased from France, taken by force from Mexico, purchased from Russia, or captured from Spain.

For anthropologists the case of the Iroquois is well-known (Abler 2007). Five "nations" of Iroquois were, and still are, spaced across present day New York State. They formed a confederacy for the purpose, they claimed, to rid themselves of feuding and warfare. It appears that first the two western nations, the Seneca and the Cayuga, joined together as did as the Oneida and Mohawk in the east. Probably the western nations next joined the central nation, the Onondaga; later the eastern nations joined to form the League of the Iroquois or The Five Nations. They might have also joined one by one. In any case, by 1600 the confederacy was operating as a government in Onondaga. Although they warred with and defeated the Huron confederacy to the north, they never achieved statehood. Consensus, not coercion, in internal affairs ruled. However, to their south, Cherokee villages moved toward statehood as they were forced to control the raiding of colonial settlements, raids that brought retaliation from the settlers. This was accomplished by incorporating war leaders into councils. Statehood followed. As anthropologist Fred Gearing stated, "the rise of the Cherokee state in the mid-1700s was one instance of independent face-to-face communities voluntarily becoming a state" (1962:106).

Two Case Studies

In this section I present in some detail two more case studies of fusion. The cultures involved are unrelated to each other, but the situations are strikingly similar. Yet, one did not achieve statehood and the other did. The first culture, the Chibuk of northeast Nigeria, near neighbors of the Higi, are nearly unknown both to the world and to anthropology. The other case is well-known to millions of people. Ancient Israel, about 1000 BCE, is intensely researched by scholars of the Hebrew Bible. In my account I have been guided by one scholar, Megan Bishop Moore, who has an interest in the warfare of the period. In comparing the Chibuk with Ancient Israel, I am conducting what anthropologists often call a controlled comparison. Many conditions are similar, but the institutions of interest, in this case the government, differ. The Chibuk did not achieve statehood; the Israelites did.

The Chibuk Hills form a chain of rock outcroppings several miles in length broken by Chibuk Mountain. Probably it was a refuge area for several hundred years before pacification by the British in the early 20th century. The total area probably does not exceed 80 square miles. Twenty-two separate patrilineal and patrilocal clan settlements arrived over time in 18 remembered migrations from many of the surrounding areas. The order of the people arriving and their names were collected by Gerald and Lois Neher (n.d.) in the 1960s. People came from Biu, Borno, Hona/Kilba, and Marghi/Higi areas. Yet they spoke one language, Chibuk, which has probably four dialects (Neher and Neher n.d.). They came seeking new farmland; they were withdrawing from feuds and intergroup hostilities; they came looking for game to hunt (elephants once were in this area); or they were robbers, chased into hiding.

The first migration settled high on Chibuk Mountain, and its headman always served as priest of the mountain's spirit. Other settlements were in and around the hills. The clan settlements consisted of perhaps 30 to over 100 compounds for a total population of 7,000 to 8,000. Each settlement was politically independent with its own headman, and to emphasize this, boundaries between were marked with tall rows of cactus hedges. Trespassing was so serious that an adult on the "wrong side of the fence" could be killed. A child could be taken as a slave. People had great suspicion of anyone from another clan. Compounds had multiple exits and no one would repeatedly use the same path for fear of ambush. Raiders, thieves, and vengeance seekers might ambush or attack at any time (Cohen 1981).

Headmen from all clans could meet to unify and defend the entire Chibuk area from an external aggressor, but they never unified to go on the offensive. However, sometimes two or three clan settlements formed an alliance for the purpose of intermarriage and mutual aid.

The Chibuk settlements were hidden in the forests and rocky hillsides of their mountainous habitat, which had springs and small streams that flowed down to the surrounding savanna. Their hills were near a major trade route from Kano to Maiduguri to Yola; the first part of the route ran from west to east, the second part, from north to south.

The Chibuk raided caravans and surrounding villages out to a radius of about 30 miles. They attacked Bura, Kilba, and Hona. The major tactic was to ambush a party, kill the men, seize the women and children as slaves, and take all the property (Cohen 1981).

With access to hidden cave retreats and secret water supplies they could withstand raids and sieges from the Borno state (Kanuri) to the north in Maiduguri. Bows and arrows and muzzle loading Dane guns, also used in hunting, kept attackers at bay. Dane guns were manufactured in the UK (Birmingham) and Belgium (Leige) (Headrick 1981:106). Arrows were poisoned (Neher and Neher n.d.) The Chibuk were not overrun by their state-level attackers. Apparently, though, other people at the same level of sociopolitical organization conducted raids on the Chibuk, killing some and occasionally burning villages.

This case study allows us to examine several issues. The Chibuk are an example of an aggregate of tribes that did not become a state. There are two hierarchical levels present because there is a primary headman, who is a spiritual leader, and a council of clan headmen. Yet the clans are political communities, which can enter into alliances with each other. Pressure from attacks might have created cohesiveness and "pushed" the Chibuk toward statehood. But this did not occur. The idea of statehood was certainly known to everyone, but no one apparently attempted to create a kingdom. I believe, because no one wanted to—there was no reason. Furthermore, any attempt by the primary headman would have been resisted by the other headmen. Given the nature of the terrain, no alliance of two or three clans could have conquered any other clan. Here is a circumscribed area. The Chibuk are surrounded by hostile neighbors. The conflict-cohesion hypothesis suggests that greater cohesion should have arisen, but it did not. Another theory that fails is the circumscription hypothesis, the argument that within a circumscribed area, increasing population may lead to one group conquering another. Thus, these explanations supporting the conquest theory of the state must be rejected.

Other mountainous areas on this earth are occupied by warring peoples who fight each other and anyone who gets near them. Examples are Northern Luzon, home of the Ifugao and Kalinga studied by Roy Franklin Barton (1919, 1949); the area of the Balkans occupied by the Montenegrins studied by Christopher Boehm (1983, 1984; Otterbein 1985b); northeastern Afghanistan, home of such cultures as the Kohistani studied by Lincoln Keiser (1991); and highland New Guinea occupied by the Dani studied by Karl Heider (1979). In none of these areas did states form. Our next case study is of peoples and the state they created, which is also located in a hilly region. Our question becomes, why and how did Israel become a state through a process of fusion?

The Hebrew peoples and others at the beginning of the Early Iron Age (1200 BCE) occupied the hilly region of the Levant or Palestine, an area about 200 miles long and 90 miles wide across the midsection. The lower half of the region was wider and less hilly. At this time, the hills were sparsely settled by perhaps no more than 20,000 people; the population density was hardly more than one person per square mile. The area was not sharply circumscribed physically, although the highest section of the hills was a north-south zone facing west toward the Mediterranean Sea. Entry into the hills from the west, however, was easy. To the east, behind these hills, were settled farmers, remote from the main cities of the Canaanites along the coast. The farmers lived in scattered villages with rectangular stone, mud brick, and cedar-wood one- and two-story houses situated around a central area. Villages were unfortified. Raising livestock, such as sheep, goats, and cattle, may have been as important as growing crops such as wheat, barley, olives, and grapes. As in villages around the world, wealth differences manifested themselves in house size, number of fields, head of livestock, number of offspring, and material possessions. Leadership rested with the wealthier individuals. Dialects would have varied greatly, but taken as a whole they were related to Canaanite languages spoken on the coast.

The Bronze Age ended in about 1200 BCE with the arrival of the Sea or Boat People, from parts of the Aegean. As foot soldiers, they attacked the Canaanite cities, defeating their chariot corps with light throwing spears or javelins and long swords (Drews 1993). Canaanites fled into the hills.

> Archaeological remains show the destruction of a few Canaanite cities and the emergence of small agricultural villages in the highlands of Syria–Palestine around 1200. [Furthermore] the evidence ... suggests that Israel and Judah originated in groups of villagers who came together for religious and political purposes. (Kelle 2007:19)

Sea People settled on the southwestern coastal strip of Canaan and became known as Philistines. Two other Sea Peoples settled to the north. They were not defeated until they reached the Nile Delta, home of the Egyptians. In a sea battle (1129 BCE) the archers of Rameses III inflicted heavy casualties on the Sea People before they could disembark.

Just as occurred in the Chibuk Hills, I believe that the region under discussion came to be settled by similar peoples who came from different directions: the settled farmers that once lived near the coast, seminomadic peoples from arid regions to the north and east after 1200 BCE, and economically depressed peoples and refugees from the coast. The core area in the central hills, from which some peoples may have spread out, was an area occupied by the tribes of Ephraim and Manasseh. Egyptian strongholds nearly surrounded the highlands throughout the Bronze Age and into the early Iron Age. (See Isserlin 1998:61, for a map of Egyptian strongholds.) They were protecting four trade routes that ran north and south. Possibly caravans on these trade routes were attacked, just as were those near the Chibuk hills.

Descent was patrilineal and residence often patrilocal as it was and is throughout the entire Middle East. Peoples were organized into tribes, clans, and **extended family households.** Clans probably occupied small villages. Tribes sometimes warred with each other. Clan-villages within a tribe would constitute a two-level hierarchical system. I suspect the villages at first were Type B societies. Village militias were the military organizations, led by wealthy individuals who became headmen and war leaders. These are the Judges of the Hebrew Bible. Alliances between tribes were sometimes formed. Warfare gradually increased between the tribes and between tribes and Canaanite armies. As patrilineal residence became more common, Fraternal Interest Groups emerged, and raiding parties based on them caused an intensification of warfare.

The basic pattern of warfare would have been ambushes and lines. The weapons consisted of shields, spears, swords (iron had slowly replaced bronze), bows and arrows, and slings. Perhaps the most famous "duel" in history or legend took place between David of Judah (an Israelite tribe south of the core area) and Goliath, a Philistine; David killed Goliath with a sling stone. I would not classify this beginning to a battle as a duel, however, since unmatched weapons were used. Apparently, with increasing population, the Israelite tribes attempted to make their way into the fertile valleys to the west. Deborah, a prophet, and Barak, the commander, defeated a Canaanite army.

Warfare with Canaanites on the coast and Philistines further south, who were better organized with their standing armies, was a threat to the Israelites. The standing armies were equipped with chariots; the village militias were equipped with shields, swords, and spears. We should now look to the conflict-cohesion hypothesis for guidance. The Kanuri were a threat to the Chibuk, but they were driven off with poison arrows. The threat to the tribes of Israel was much greater, particularly to the southern tribes that included Judah and Benjamin.

The Hebrew Bible tells us that the leaders and elders of Benjamin asked a prophet named Samuel to appoint a king to govern them. Remember that every adult would have been familiar with the idea of statehood. Samuel selects Saul, a wealthy Benjaminite, who rules from about 1020 to 1005 BCE. A three-level hierarchical system is formed, our criterion for a Maximal Chiefdom or Inchoate Early State. The tribe of Benjamin was between Judah to the south and many of the tribes that were in the north. Saul called his kingdom Israel, which included Ephraim, Manasseh, Gilead, and Benjamin. Its population might have risen to 50,000.

According to military writer Richard A. Gabriel, Saul created a corps of full-time professional soldiers to command the militias (2003).

The nucleus of the standing army consisted of perhaps 3,000 men, the force being divided into two formations. The smaller one of about 1,000 soldiers was commanded by his eldest son, Jonathan. After winning numerous battles, a showdown with the Philistines took place at Mount Gilboa. The Philistines attacked from the north with their professional army equipped with chariots. A rival for Saul's kingship, David of Judah, aligned himself and his followers with the Philistines. Archery fire from chariots drove Saul up Mount Gilboa. Rather than be captured, he fell on his sword. Jonathan and other sons were killed (Herzog and Gichon 2006).

Saul's death opens the way for David to become king of Israel (1005 to 965 BCE). He first becomes king over Judah in the south, then over tribes in the north, and soon king over all of Israel. He wins victories over the Philistines and establishes Hebron in Judah as his capital. His army consists of his elite warriors and village militias (Herzog and Gichon 2006). Later he moves his capital to Jerusalem, just to the north of Benjamin. The conquest theory of the state can be evoked here. Clearly David had the task of uniting the tribes through coercive diplomacy and conquest. Recall that in the latter phase of the Inchoate Early State warfare flourishes.

The attacks by the Philistines on the tribal peoples of Benjamin and Manasseh seem to have created the kind of cohesiveness that was observed to develop in the Robbers Cave Experiment. A leader or king is selected. But a rival from another tribe, rather than joining with the king, aligns himself and his followers with the culturally different enemy, as David did with the Philistines. We observe a common situation whereby one faction in a polity or an entire polity surrounded by culturally similar polities, seeks help from a culturally different polity to help defeat its opponent. Polities within a culture typically do not unite to defeat a culturally different enemy (Otterbein 1968b). The Yoruba kingdoms of West Africa provided a case study. Like the Yoruba kings, Saul and David did not unite to defeat the Philistines. Rather, David allied himself with the Philistines to defeat Saul, and he was successful. Next, with state organization in hand, David began to defeat the tribes and unify the kingdom.

In primary state formation there are no other states that can provide pressure or help to one of the factions. But once primary states have formed, not only can they war with each other, factions within each can seek outside help. For Israel it may have been necessary for state formation to occur before a rival within could seek outside help. Before Saul became king it would have been tribe against tribe—feud/internal war, but no conquest possible, and no government to create consolidation.

A famous Biblical story is David's adultery with Bathsheba and how he achieved the death of her husband, Uriah. Solomon was the fruit of that union. Solomon, David's fourth son, expanded the kingdom further, built fortifications near border areas, and constructed regal buildings. David had so successfully established his kingdom that his son did not need to wage war. Solomon created administrative *districts.* With consolidation of the kingdom, a four-level polity emerged, a typical Early State: village, tribe, district, kingdom. Like David, Solomon had a long reign (965 to 915 BCE). Solomon apparently had a large chariot force that never engaged in combat (Isserlin 1998:81). (Excellent maps showing the expansion of the kingdom under each monarch are in Isserlin [1998:63–82], and if you are interested in weapons and tactics for this period in the Near East go to Yigail Yadin's *The Art of Warfare in Biblical Lands* [1963].)

Although defensive needs and personal ambition seem to be the driving forces that led to Israelite statehood, the role of religion was probably also important. Yahweh, the Israelite *high god,* seems to be one of many existing gods before the time of David. I suspect that Yahweh is promoted as the only god during David's reign. One ruler, one god. David in effect said, "You owe homage to me, you must worship my god, Yahweh." This pattern of high gods, frequently accompanying three-level polities, was found by sociologist Guy Swanson in his cross-cultural study of *The Birth of the Gods* (1960). In other words, polities organized with at least three levels are likely to have a high god. Sometimes the high god rules over superior gods, other times the high god has no other gods to rule over. He gets to focus on humankind. This is known as *monotheism.* I view the belief in a high god, who is the only god in a belief system, as a device that can be used by the ruler of a newly formed state to consolidate his power. Many such high gods, it seems to me, from a cross-cultural perspective, are punitive. I suspect this occurs because they originated in the period when a despotic ruler was establishing his kingdom. And this happened in ancient Judah 3,000 years ago. The belief in a high god in Judah and elsewhere is a means of legitimizing a newly formed state.

Chapter Eight

The Termination and Prevention of War

States, or for that matter all polities, are fragile. A government controls a state (but governments also exist in uncentralized political communities). A government can be replaced by a conquering polity, and yet the state may survive, or both government and state can be destroyed. The government may be replaced by an enemy army with the population more or less remaining in place, perhaps to continue doing the hard work of the country. The people can also be killed or removed, perhaps to become slaves in the conquering nation. Boundaries can expand and contract; one common loss in war is the loss of territory. An uprising by a segment of the population—a social class or an ethnic group—can destroy the government. Some governments are more afraid of their "own" people than they are of a neighboring, hostile polity. One type of uprising is a civil war, which can split the state into two or more polities. In small-scale societies, fissioning of polities commonly occurs, with the segments either remaining on amicable terms or becoming mortal enemies. A feud can turn to internal war in this fashion. A war can have fatal consequences for a people and their way of life. It is so important to end and prevent wars.

In a small book titled *Every War Must End* (1991), Fred Iklé, a political scientist and former Undersecretary of Defense for Policy in the Reagan administration, described how wars quickly get out of the control of either side and usually end abominably for one or both sides. Although Iklé wrote about state-level warfare in the 20th century, his general findings pertain to prehistorically known polities and probably also to uncentralized political systems. In this last chapter I will provide some thoughts as to what can happen to peoples who go to war.

These are conclusions drawn from the analyses and case studies presented in chapters 1–7. I will emphasize the dire consequences that can arise from initiating war, not only from being attacked.

The outcomes of war are likely to be the same or similar for peoples at all levels of sociopolitical complexity and at any time since the rise of Homo sapiens.

CAUSES AND OUTCOMES OF WAR

Strategies and Tactics

Our two best known examples of warring nonliterate societies, the Dani and Yanomamö provide evidence of how alliances are formed to give a polity a military advantage; the alliance that strikes first may succeed in annihilating another polity. Competition over scarce resources (CSR) has been argued to be the cause of this type of warfare. Nonetheless, the Dani and the Yanomamö examples suggest that defense and revenge are more important. Indeed, their warfare seems to be about polity survival.

Uncentralized political systems have developed a basic pattern of combat that combines lines and ambushes; the purpose of using two types of tactics is to increase the chance of success. Line battles are more likely to test the strength and will of each side rather than to be an opportunity to display feather headdresses and new dance steps. Ambushes before or after a line battle often have a devastating effect on those surprised, particularly if they were asleep at the time of the attack.

Centralized political systems are more likely to have a basic pattern composed of lines and siege operations. With the rise of fortifications to protect villagers or city dwellers from surprise attack, military organizations intent on destroying villages and cities develop siege operations that can destroy walls and fortifications. Perhaps the most effective weapons for city destruction are the artillery shell and the bomb. The most destructive weapon is the atomic bomb. Ambushes of sleeping villages, siege operations of cities, and shelling and bombing attacks of cities all invariably kill many noncombatants.

Although casualties in combat can be low, they are often high. A "successful" ambush almost by definition means nearly all those attacked are killed, whether it be a patrol, an outpost, or village. As weapons became perfected, and a shift took pace from projectile to shock weapons, casualties increased. Engagement in hand-to-hand combat, as with Greek hoplite armies, often resulted in high casualties. The killing does not usually stop when a battle is over. Fraternal Interest Group based societies and despotic states usually kill captured enemy warriors and frequently the old men, women, and children of a village, town, or city.

Reasons for War

While nearly all polities at one time or another become embroiled in war, as aggressor or victim of attack, two of four major categories of polities and their military organizations are much more likely to be aggressors: The Type A Society and the Early State. These are polities with Fraternal Interest Group based military organizations and despotic states with elite warriors who are the supporters of despots. In recent years it has been the countries with Fraternal Interest Groups and strong notions of honor and revenge, as well as newly formed dictatorships, that are most likely to attack their neighbors. When both features are combined in one society, as is true for many African and Middle Eastern countries, war seems almost inevitable.

Some decisions to go to war are based on the diverse interests of decision makers who can easily plan and launch an attack but give little or no thought to how the combat may end. This goals-of-war approach contrasts with the usual scholarly analysis of the causes of war, which typically focuses on *underlying* causes. The need to obtain scarce resources is most often seen as the underlying cause, but scarcity is a matter of perception. Resources within a country may be seen to be scarce only because people and their leaders *want* these resources—not because they are necessary. These wants are manifest in the decisions of some leaders. There may be, however, better ways of meeting these wants than going to war. The link between material and efficient causes needs far more investigation.

To obtain scarce resources some polities (rather their leaders) may believe it is necessary to seize territory occupied or claimed by other polities. The modern conception of territorial boundaries of states seems to lead to conflict. A nation may not need the land now but may want it for future expansion. With future needs in mind a leader(s) may embark on seizing land. The expansion of the United States in the 19th century is an example.

Ethnocentrism itself shapes warfare and makes the outcome unpleasant for the losers. Enemies vilify each other. If the enemy is seen as an "animal," one with undesirable characteristics, killing, torturing, and enslaving come easier to the victor. Ethnocentrism contributes to the killing of captured enemies. Furthermore, ethnocentrism is taught and encouraged to young males who are being socialized informally and formally to be warriors or soldiers. Socialization for war plays an important role in societies that plan for war and expect to be engaged in frequent combat.

The Evolution of War

Hunter-gatherers are of great interest to both laypersons and social scientists. I feel that I have learned much about the causes of war from my study of hunter-gatherers, and this I was able to do by avoiding adherence to a specific ideology. If hunter-gatherers do *not* engage in war, we know that war is not in our genes (Fry 2006, 2007); if all hunter-gatherers engage in war, we surmise that war is in our genes. If we believe war is in our genes, then we conclude that war began more than two million years ago (LeBlanc 2003). My conclusion—that some hunter-gatherers engage in war and others do not—is a challenge to both views. Both social organization (Fraternal Interest Groups) and subsistence base (big game hunting, whether terrestrial or aquatic) seem to be responsible for the armed combat of four types of hunter-gatherers. The four types of hunter-gatherers who do not engage in war do not have Fraternal Interest Groups and do not hunt big game. It seems to me that the lesson for us is that material causes, rather than genetics, are responsible for war. Even when conditions are ripe for war, triggers in the form of efficient causes (individuals around the table or camp fire) are required to initiate raids and battles.

The study of primary states also contains lessons. The development of inequality in agricultural villages led to intense rivalries that could result in homicidal struggles. The winners became a centralized government that developed the idea of the state. With a strong military and a surplus to support it, an Early State was able to conquer other Early States. If warfare had occurred early in the developmental cycle, domestication of plants would not have taken place and states would not have arisen. Ironically it was not war but internal conflict that led to the conquest state. These Early States were also despotic.

The study of secondary states brings forth similar lessons, whether the states arose through fissioning or fusion. Warfare can lead to the breakup of the state, or warfare can lead separate polities to coalesce into a single polity. In either situation, the process is likely to be violent. Political assassinations occur with frequency whether a state is breaking up (here a ruler may be killed by a close associate, shot from a distance, or seized and incarcerated) or several states are joining together under one ruler (the heads of rival polities are often executed). Whether long ago or today, state formation is usually a series of bloody events.

Although it can be argued that the entire world is a warfare system, some regions, both ethnographically, like highland New Guinea, or contemporaneously, like the Middle East, can be analyzed as a warfare system. If the polities are culturally different, as in the Middle East, change will come rapidly. With rapid change polities may break up and new, smaller polities form; sometimes one polity conquers or joins another. In all of these situations large numbers of people may die, particularly noncombatants who are less able to flee and will probably be unarmed.

Many studies have tried to describe the root causes of war, and how to dig them out and destroy them (Brown 1987). Later in this chapter, I will set forth some things that governments, institutions, social groups, and individuals can do to curb war. However, I realize the *future is unknowable and uncontrollable.* An unexpected event can occur—a **black swan**—that totally alters events (Taleb 2007). I also realize that the *remembered past* is *the present.* Presentism affects all historical writings, and it has affected my writing of *The Anthropology of War.* We back into the future. According to Robert Persig in *Zen and the Art of Motorcycle Maintenance,* the Ancient Greeks held the following view of time:

> They saw the future as something that came upon them from behind their backs with the past receding away before their eyes. . . . All you can do is project from the past, even when the past shows that such projections are often wrong. And who can really forget the past? What else is there to know? . . . What sort of future is coming up from behind I don't really know. But the past, spread out ahead, dominates everything in sight. (1974:413)

Is it an oversimplification to say that we back into the future? Lendon (2005) describes how the Greeks, though influenced by Iliad-style warfare of the past, backed into modern warfare with its Integrated Tactical System.

Military thinkers attempt to devise means that can keep their polities viable. Credence has been given by some to two approaches: deterrence and retaliation. I will critique both in the next section. Deterrence requires the building of a defensive military organization that is *perceived* by one's neighbors as capable of repelling an attack, while retaliation requires the building of a military organization that has the capacity to destroy an attacking force and inflict damage to the aggressors commensurate with the damage done by them.

THE DILEMMA OF DISARMING—AND ARMING

The "dilemma of disarming" is an expression that I use for the dilemma an individual or nation finds itself in when it needs to decide whether to disarm itself or to choose the alternative, that is, acquire arms for protection. To disarm in a neighborhood of hostile polities is to risk being taken over, with death to the leaders and enslavement of the population. Thus, it is better to arm.

> While the possession of armies and weapons may prevent the annihilation of a nation and give gun owners the means to kill assail-

ants, it also gives them the ability to carry out, for reasons of aggrandizement (and sometimes just to test the weapons), attacks upon other nations and individuals. Not only may they be tempted to use the arms in non-self-defense situations, but their mere possession [of the arms] may provoke others to attack because they either fear being attacked or desire to take the weapons for their own use. Hereupon lies the dilemma of disarming. Weapons are needed for survival, yet their possession is likely to involve their owners in the very armed conflicts which the weapons were intended to prevent. Ironically, it is the basic need for self-preservation— as a nation or as an individual—which is both the reason for arms possession and the cause of the killing which the weapons were intended to prevent. (Otterbein 1989a:128–129)

Individuals carry weapons, whether knives or handguns, for self-defense, while a polity builds a military organization of warriors or soldiers who carry spears or firearms. Whether the arms bearer is an individual or an army, either can defend or can attack. Defense is the essence of deterrence, attack is the essence of retaliation.

Deterrence

Deterrence theory is a doctrine developed by defense planners. It has become the official position of nearly all modern nations—the rationale for building a strong military organization. "If we have an excellent military, neighbors will be afraid to attack us." Furthermore, an excellent military organization is seen by the nation "investing" in such an institution as a source of pride and respect. The theory behind the deterrence doctrine is that "military readiness prevents an attack by a potential aggressor by creating in the minds of its political and military leaders the belief that if they were to attack, their military would be defeated" (Otterbein 1989a:129). The key notion in the theory is perception. If your enemies think your military is weaker than theirs, deterrence may fail; if they believe they can defeat you, they are likely to attack. The most famous example from my generation is the Japanese attack on Pearl Harbor in the Hawaiian Islands on December 7, 1941. By April 18, 1942, United States B-25s had bombed Tokyo, and on June 4, 1942, at the Battle of Midway, the United States Navy decisively defeated the Japanese Navy.

The doctrine of deterrence developed during the last half of the 20th century after the first atomic bombs were exploded at Hiroshima and Nagasaki on August 6 and 9, 1945. The rationale for building nuclear weapons has been that "if we have the bomb, no one will dare attack us." Thinking about deterrence has thus been part and parcel of the atomic age. The eagerness that some polities have for acquiring *the bomb* seems based on this notion. While preparing my cross-cultural study of capital punishment I encountered in the writings of anthropologist Edmund Leach the assertion that an explicit theory of general deterrence is inherent in all legal systems. In Leach's view, legal systems are based at least in part on the assumption that potential wrongdoers are inhibited from breaking rules by the fear of consequences that would follow, and that the "actions of judges and policemen are based on this premise" (1977:31). I found no data to support this claim (1986).

Several cross-cultural and cross-historical tests of deterrence theory have failed to find support for the theory. In my study of the evolution of war, I did not find that well-prepared polities with high military sophistication scale scores were attacked less often, and these polities were more likely to attack their neighbors (1970). Raoul Naroll, historian and comparative anthropologist, obtained similar negative results, first in a cross-cultural study (1966), then in a cross-historical study (Naroll, Bullough, and Naroll 1974). A positive result that Naroll found in the cross-historical study also seems to pertain to nonliterate societies. Polities that had water transport, meaning that warriors or soldiers can be moved rapidly over great distances, like the Vikings, were engaged in warfare more frequently than landlocked nations (1974). The warring peoples of the American Northwest Coast had ocean-going war canoes that transported attacking forces.

In a study of 20th century warfare, political scientist John Mearsheimer also found that military preparedness led to war. He focused on "conventional deterrence" (non-nuclear deterrence). However, the nations in his sample seemed uninterested in defense. Rather, they developed military armament programs based on the German concept of the blitzkrieg ("lightning war"). Such a military force is composed of tanks, armored vehicles, and dive bombers. Being fearful of their neighbors, his sample nations built military organizations that they planned to use to attack neighbors before their neighbors could prepare a defense against the attack. They were usually successful in their use of the blitzkrieg (Mearsheimer1983).

It appears that military planners use deterrence theory to justify arms build up, arms that they may plan in advance to use to attack neighbors. Add a naval transport system, and a blitzkrieg army can be taken any place, up or down the coast, as on the Northwest Coast, or across the oceans, as did the Spanish conquistadors with their horses, firearms, and body armor. Cortes defeated the Aztecs and seized their capital, Tenochtitlan, in August 1521 (Hassig 1988).

Retaliation

Deterrence theory has recently been replaced by thinking that the best way to deal with aggressors is to punish them. If they attack you, you attack them. Deterrence thinking, however, lies hidden in retaliation theory; it is believed that the punished aggressor will not attack a second time. Both Douglas Fry and Azar Gat, one of whom sees peace many places and the other only war, have argued that a computer-derived model has shown that the best strategy is not to attack or retreat, but to retaliate. They both base their reasoning on computer simulations by game theorist Robert Axelrod, who found a tit-for-tat strategy as the way to deal with aggressors best. The logic is as follows (Daly and Wilson 1988:235, ital removed): "Cooperate the first time, and then match your opponent's last move." If your opponent cooperates and you continue to reciprocate, conflict or war does not arise. If he defects, so do you. A stalemate presumably arises. Not to retaliate could lead to defeat. To overretaliate could lead to a suicidal escalating war. Fry also uses the arguments of J. M. Smith and G. R. Price who used computer simulations to compare the relative success of different fighting strategies (Fry 2006, 2007; Gat 2006). The approach, I believe, is flawed. It assumes no first or preemptive attack and the exact same response to an attack, neither of which occurs in practice.

My critique of retaliation theory is based on two problems: (1) In order to be able to retaliate, a military buildup must occur. Even in small-scale societies the men must be able to use weapons efficiently. If the enemy sees new weapons being made or obtained in trade, plus men training with these weapons, they may launch a preemptive attack, a raid. (2) Retaliation may lead the attacker to counterattack, this time more forcefully. Thus, escalation of conflict occurs. Attack and counterattack may go back and forth. For an example, see Yanomamö levels of armed combat. The outcome of an escalation may be the defeat and destruction of one of the polities.

Where does this leave the strategic thinker? Fry is optimistic; he believes that people can cooperate. Gat is pessimistic; he believes there are no good strategies. For Fry, interpolity conflict can be reduced; for Gat, warfare seems to be everywhere, war being inevitable. My views follow.

PREVENTION OF WAR

Thus far in this last chapter, I have stressed both the fragility of states and the causes and outcomes of war. I have also briefly discussed military strategies to prevent war, namely deterrence and retaliation; and I have described why these twin approaches rarely accomplish what their creators intended. Either approach is like shooting a pistol or rifle in self-defense and having the weapon blow up in your face. In this final section I will be making a number of suggestions drawn from what we have learned from the anthropology of war, which may be helpful in preventing or at least ameliorating the horrors of war.

Alliances made with other polities both near and far are important for polity survival. I see two types of alliances—military, usually
called defensive alliances, and economic, which involve trade, educa-
tional exchange, scientific cooperation, and the like. I see student
exchanges involving language and cultural learning as extremely
important. However, learning the scientific knowledge of another
nation can at times be spying. Military alliances should be joined only
for defensive purposes among neighbors as a way to promote peace.
Military alliances with distant nations can embroil the nation in war.
President George Washington, in his farewell address to the nation,
published September 19, 1796, said to avoid entangling alliances
(Flexner 1984:151–153).

On the other hand, economic alliances with faraway nations are
of great importance. For one thing, they bring us exotic goods and foods
like Japanese hybrid cars and sushi or German motor cars and beer.
But as an aging anthropologist I want to see legions of young anthropologists visiting foreign lands, learning the peoples' cultures, and
bringing this knowledge back to the United States. I also want to see
the reverse, with students from other lands coming here, learning our
culture, and returning to their countries. Finally, I would like to see
peoples of different nations working together on Superordinate Goals,
trying to devise, for example, means to lessen the effects of earth-
quakes, hurricanes, tsunamis, and epidemics. We should be battling
these things, not each other.

Avoid excessive arms buildup for it can lead to preemptive
attacks and arms races. Certainly, a nation needs armed forces, but
these should be designed for defense against foreign forces, insurgency,
terrorism, and domestic problems. It is not true that "the best defense
is a good offense." Building blitzkrieg-style armed forces, on the other
hand, is the first step to war. Selling these types of weapons to other
polities certainly does not stabilize those areas of the world. The Swiss
with their armed citizenry (militia) and multiple fortresses built into
the mountains (World War II) had a military sufficient to keep the Germans and Italians from invading. The Swiss Army of 500,000 men
comprised 40 percent of the male population. Bolt-action military
rifles and ammunition were kept in their homes (Richter 1943). The
United States' Second Amendment about militias and the right to bear
arms seems to refer *both* to keeping individual weapons at home as
well as to the militia that those armed citizens would form if the nation
were attacked.

We have seen in this volume that the sociopolitical organization of
a polity, and the type of military organization that is compatible with
it, are factors in the likelihood of war. The figure below (which also
appears in chapter 3) summarizes the systems in which warfare is common and uncommon:

	Warfare Common	Warfare Uncommon
Centralized Political Systems	Early/Despotic State (Elite Warriors)	Mature State (Conscripts)
Uncentralized Political Systems	Type A Societies (Fraternal Interest Groups)	Type B Societies (Village Militia)

Figure 8.1   Types of Sociopolitical Organization Related to War and Peace

In terms of world peace, it would be desirable if polities in the left column could move toward the right column. There is, however, little that can be done to make them do so. Nevertheless, and unfortunately, polities in the right column can be forced toward the left column if they are attacked. This leads to defensive external war. Recall that I put such polities in the category I once reserved only for societies without military organizations and war.

Early/Despotic States over time slowly "mature." It appears that it takes about 200 years for a Mature State to develop. Waiting for this to happen may be the best strategy for maintaining peace, but an impossible one since actions that can be viewed as military provocations may be what the despot wants. The conflict/cohesion hypothesis is well-known to despotic rulers. Any attack on his polity is the basis for tightening the reins of control over his government. Perhaps it is best to "pleasure" the despot. Provide him with a mega mansion, give him access to Western movies, rock stars, and escort services. Cultures with ruling elites composed of related males can be bellicose. If ways can be found to get some of them out of their country, this might help. I suggest jobs in the arts and education.

Countries should strive for self-sufficiency in products and goods that are essential for survival. Food, water, and energy have been in the news in recent years. If a nation needs what another country has, this could lead to aggressive war—offensive external war. If a nation does not need what the other has, these reasons for attacking are not present. Since clean water is essential for life, this is the single most important natural resource that a country must have and preserve. The Great Lakes between Canada and the United States hold approximately 20 percent of the world's fresh water. Perhaps some day we can gather up the world's weapons by exchanging them for fresh water—like a gun buy-back.

A polity and its people should curb or resist chauvinism and ethnocentrism. Governments should avoid propaganda campaigns that vilify neighbors. Promoting the virtue of its own polity and people without making negative comparisons to others can be a major ingredient in laying the ground for avoiding future war. Do not glorify war. Glorifying battles almost always puts the enemy in unfavorable light: the "sneaky bastards" used a surprise attack. Battlefields, however, can be used to highlight the service, courage, and compassion of both armies. This has been done for U.S. Civil War battles and for battles in the War of 1812, between Americans and British Canadians. Peace should be glorified. The notion that peace is the absence of war should be replaced with the idea of Positive Peace—promoting Peace as a goal of human existence.

Indeed, the war trials at the end of World War II, through their charges, have laid out what is inappropriate conduct in international affairs: (1) aggressive warfare, (2) conspiracy, (3) war crimes, and (4) crimes against humanity. International organizations and agencies, as well as the United Nations, have established norms of conduct for nations. Stated like commandments, the norms are as follows: (1) Do not attack another polity. (2) Do not conspire with other polities to attack another polity. (3) Do not torture and kill captured enemy soldiers. (4) Do not target and kill noncombatants. All nations should adhere to appropriate international conduct and norms for nations.

Finally, follow the Golden Rule in international affairs. "Do unto others as you would have them do unto you." Ironically, this is consistent with retaliation theory. Or, colloquially, "Don't mess with others."

Appendix

---

40 Warring Peoples

The following list of 40 cultures includes some of the best known, best described, or famous people known to anthropologists. The list is not intended to be inclusive. All have been described or mentioned in this book. Hence, each is accompanied by one or two source citations that are listed in the Bibliography of this text. These sources provide a more in-depth look at these cultures and contain references that readers can use to find additional information. The list of sources below is not a great books list, a list of best sellers, or any kind of ranking such as exists for novels, movies, or TV shows. Instead, it is a way to get you started on your journey to discovery.

Andamanese—Kelly, 2000:75–119
Arunta/Aranda—Spencer and Gillen, 1927; Murdock 1934:20–47
Aztec—Hassig, 1988; Kurtz, 1978
Bellona—Kuschel, 1988
Bushmen—Thomas, 1959
Central Eskimo—Boas, 1888
Chavin/Moche—Otterbein, 2004:130–142
Cherokee—Gearing, 1962
Comanche—Hoebel, 1940, 1954
Cheyenne—Hoebel, 1978
Dani—Heider, 1970, 1997
Greek/Hoplite—Lynn, 2003
Greek/Iliad—Lendon, 2005:15–161
Iban—Vayda, 1976
Ifugao—Barton, 1919
Iroquois—Morgan, 1962[1851]; Otterbein, 1979b
Israel (Ancient)—Yadin, 1963; Gabriel, 2003; Kelle, 2007
Jivaro—Bennett Ross, 1984

Kalinga—Barton, 1949
Kapsiki/Higi—Otterbein, 1968a; van Beek, 1987
Kentucky Feudist—Rice, 1978; Otterbein, 2000b
Kwakiutl—Benedict, 1934:173–222; Codere, 1950
Kohistani—Keiser, 1991
Macedonians—Adcock, 1962; Engles, 1978
Maori—Vayda, 1976
Maring—Rappaport, 1967; Vayda, 1974
Montenegro—Boehm, 1983, 1984
Murngin—Warner, 1931, 1937
Nuer/Dinka—Kelly 1985; Otterbein, 1995, 2004:203–207
Pygmies—Putnam, 1948; Service, 1979
Semai—Dentan, 1979[1968], 1992
Shang/Chinese—Otterbein, 2004:158–173
Tiwi—Hart and Pilling, 1960
Tonga—Michener and Day, 1958
Uruk/Sumer—Otterbein, 2004:142–158
Waorani—Robarchek and Robarchek, 1998
Yanomamö—Chagnon, 1968; Ferguson, 1995
Yoroba—Ajayi and Smith, 1964 (See Otterbein 1966 for reference)
Zapotec—Marcus and Flannery, 1996
Zulu—Ritter, 1957; Edgerton, 1988

Glossary

alliances—*See* alliance system.

alliance system—A warfare system in which the political communities have entered into alliances. It is the alliances that wage war against each other.

armed combat—Fighting with weapons.

bilateral—*See* kindred.

black swan—A highly improbable event with three principal characteristics: it is unpredictable; it carries a massive impact; and, after the fact, we concoct an explanation that makes it appear less random, and more predictable, than it was.

chief—A formal political leader with limited power.

clan—A descent group composed of two or more lineages. Clans are nonlocalized, they are named, and they are based on the same rule of descent as the lineages that compose them.

closed warfare system—A warfare system in which the armed combat is between culturally similar polities.

coercive diplomacy—Threatening the other side with military attack. A tactic used by a militarily strong state when negotiating with a weaker state. The weaker state is expected to give up land, resources, or other possessions desired by the stronger state.

compensation payments—Feuding with compensation payments occurs if the relatives of the deceased sometimes accept compensation in lieu of blood revenge. Feuding without compensation occurs if the relatives of the deceased are expected to take revenge through killing the offender or any close relative of the offender. The possibility of paying compensation does not exist.

conflict/cohesion hypothesis—A theory that when there is conflict between groups, there is cooperation within the groups that are in conflict with each other.

controlled comparison—This research technique requires that two highly similar cultures be compared simultaneously. Factors or culture traits in common are "controlled," in that they do not influence what the researcher

113

is trying to explain. For example, Waorani warfare (Semai and Waorani compared) and the rise of ancient Israel as a state (Chibuk and Israel compared). Only differences between the cultures are explained.

cross-cultural study—A research technique that tests hypotheses by comparing simultaneously information from a large number of cultures, perhaps 20 to 100. One or more dimensions or variables, with points or categories on each, are coded for each culture.

culture—The way of life of a particular group of people. Culture includes everything that a group of people think, and say, and do, and make.

culture, a—A particular group of people who share the same way of life. It is usually an ethnic unit composed of contiguous political communities that are culturally similar. The criterion for distinguishing one group of people from another is language. If two groups of people speak different languages, they are different cultures.

dictator—An informal political leader with absolute power.

diffusion—The geographic spread of culture traits. It can occur through the acceptance of culture traits from one culture by the members of another culture, or it can occur through the migration or dispersion of local groups belonging to one culture.

dispersed homesteads—Settlement pattern in which the members of the culture reside in permanent homesteads with fields between the homesteads.

district—A territorial unit within a political community that is composed of local groups.

dueling—Armed combat between two persons fought with matched lethal weapons under agreed on conditions. The duelists are from the same social class with honor being the most frequent motive.

ethnocentrism—"The technical name for [the] view of things in which one's own group is the center of everything, and all others are scaled and rated with reference to it" (Sumner 1906:13).

ethnographer—An anthropologist who conducts research on a culture of a particular people and who writes a description of their culture.

ethnography—A descriptive account of a people's culture. Usually an ethnography describes the culture of a particular local group within the culture.

exogamy—The practice of marrying an individual who is not a member of one's descent group. It is marriage outside either a lineage or clan.

extended family household—A domestic group consisting of three (or possibly more) generations of related individuals.

external war—Warfare between political communities that are culturally different. There are two types of external war: (1) political communities of a cultural unit can either attack or (2) be attacked by culturally different political communities.

feuding—A type of armed combat occurring within a political community in which, if a homicide occurs, the kin of the deceased take revenge through killing the offender or any member of his kin group. The pattern of revenge may continue for many years.

Fraternal Interest Groups—Localized groups of related males who can resort to aggressive measures when the interests of their members are threatened. Found in Type A societies.

goals of war—The reasons that military organizations go to war—subjugation and tribute, land, plunder, trophies and honors, revenge, and defense.

headman—An informal leader with limited power.

hierarchical level—An administrative level within a political community. Local groups constitute one level; if local groups are organized into districts this is another level; if districts are organized into provinces this is a third level. Either districts or provinces can be organized into a state or kingdom.

high god—A supernatural being who created the universe and/or is the ultimate governor of the universe.

horticulturalists—People who practice farming or raise crops with the use of hand tools, such as a digging-stick or hoe.

household—A group of people living together who form a domestic unit. Usually the group occupies a physical structure with walls and a roof, which can be described as a house or homestead.

hunting and gathering—Including fishing, consists of techniques of obtaining natural foodstuffs—animal and vegetable—from the environment.

hypothesis—A statement of a precise relationship between that which is to be explained and that which is to do the explaining. The relationship is a linkage between two culture traits or two concepts that the anthropologist wishes to test in order to determine the validity of the relationship.

internal war—Warfare between political communities within the same culture.

kindreds—An aggregate of kinsmen that consists of those relatives of an individual who have identical rights and obligations with regard to the individual. Such kindreds need not be named, but probably are. They are said to be bilateral.

king—A formal political leader with absolute power.

lineage—Either a patrilineage or a matrilineage.

local group—A spatially distinguishable aggregate of people. It my be as small as a single family or as large as a city.

matrilineage—A descent group whose membership is based on a rule of matrilineal descent.

matrilineal descent—A cultural principle that automatically filiates a child at birth through his or her mother to a descent group that consists of all kinsmen who are related to the child through female ancestors.

matrilocal residence—Marital residence rule that requires the groom to reside with the bride, either nearby or in the home of the bride's parents. Often referred to as uxorilocal residence.

maximal territorial unit—An alternate term for political community.

military organization—A type of social organization that engages in armed combat with other similar organizations in order to obtain certain goals.

monotheism—Belief in one god. If a high god is present, but superior gods are not, the religious belief system can be classified as monotheism.

multilocal—*See* multilocal residence.

multilocal residence—Marital residence in which the bride and groom establish a household either nearby or in the home either of the bride's parents or groom's parents. Also referred to as ambilocal residence.

Neolithic—The New Stone Age—began about 8000 BCE in the Old World. Polished stone tools, food production, and pottery characterize the period.

nomadic—Settlement pattern in that the members of the culture are grouped into small bands that shift from one section of their territory to another throughout the year.

nonprofessional military organizations—Composed of military personnel who have not had intensive training in the art of war.

open warfare system—A warfare system in which the armed combat is between culturally different polities.

Paleolithic—The Old Stone Age—began about 2.5 million years ago and ended about 10,000 years ago or later. The Middle Paleolithic, or Middle Stone Age, began about 180,000 years ago. The Upper Paleolithic, or Late Stone Age, began about 40,000 years ago.

patrilineage—A descent group whose membership is based on a rule of patrilineal descent.

patrilineal—*See* patrilineal descent.

patrilineal descent—A cultural principle that automatically filiates a child at birth through his or her father to a descent group that consists of all kinsmen who are related to the child through male ancestors.

patrilocal—*See* patrilocal residence.

patrilocal residence—A marital residence rule that requires the bride to reside with the groom, either nearby or in the home of the groom's parents. Extensive hunting and fishing are related to patrilocal residence. Often referred to as virilocal residence.

political community—"A group of people whose membership is defined in terms of occupancy of a common territory and who have an official with the special function of announcing group decisions—a function exercised at least once a year" (Naroll 1964:268). There is usually more than one political community in a culture, but it is a maximal territorial unit—it is not included within a larger unit.

political leader—The leader, official, or head of a political community.

political system—The organization of hierarchical levels within a political community.

polity—A commonly used term for a maximal political entity. A polity is politically independent of all other polities. The term is used in this text as a synonym with political community.

polygynous—*See* polygyny.

polygyny—The marriage of one man to two or more women at the same time.

population density—The number of persons per square mile. The formula for computing is: (Population Size) divided by (Square Miles).

Positive Peace—The belief that war is preventable.

professional military organizations—Composed of military personnel who devote a substantial part of their time during their early adulthood to intensive training, which may involve not only practice in the use of weapons but also practice in performing maneuvers. They may be members of age-grades, military societies, or standing armies.

redistribution—The systematic movement of goods and services toward an administrative center and their reallocation by the authorities.

ritual warfare—The mistaken belief of social scientists that the warfare of peoples living in uncentralized political systems is viewed by the participants as nonserious, involving few casualties; as rule bound; and as a game.

Two lines of warriors face each other and throw spears or shoot arrows. Projectiles fly until someone is wounded. The fighting stops (Otterbein 2004).

self-redress—One disputant takes unilateral action in an attempt to prevail in a dispute or to punish another. Also called self-help, coercion, or folk justice.

slash-and-burn cultivation—A type of horticulture in which the vegetation removed from the land is heaped and burned, and the ash is used to fertilize the soil.

state—A political community headed by a king or dictator.

tactics—The "science" of placing and maneuvering military organizations for combat. Leaders develop a plan of attack (ambush, line, or siege operation) or a plan of defense (fortifications).

uxorilocal residence—*See* matrilocal residence.

virilocal residence—*See* patrilocal residence.

warfare—Armed combat between political communities.

warfare system—The warfare that occurs over a period of time between or among two or more rival political communities or alliances.

weapon—Any instrument used in combat. Weapons are classified as either shock or projectile.

Bibliography

Abler, Thomas. 2007. "Iroquois: The Tree of Peace and the War Kettle." In *Discovering Anthropology: Researchers at Work*. Carol R. Ember, Melvin Ember, and Peter N. Peregrine (eds.). Upper Saddle River, NJ: Pearson/ Prentice-Hall.

Adcock, F. E. 1962. *The Greek and Macedonian Art of War*. Berkeley: University of California Press.

Ardrey, Robert. 1966. *The Territorial Imperative*. New York: Atheneum.

Aristotle. 1984. *The Politics*. Carnes Lord, trans. Chicago: University of Chicago Press.

Ayers, Barbara. 1974. "Bride Theft and Raiding for Wives in Cross-Cultural Perspective." *Anthropological Quarterly* 47:238–253.

Bailey, F. G. 1969. *Stratagems and Spoils: A Social Anthropology of Politics*. New York: Schocken Books Inc.

Barnard, Alan (ed.). 2004. *Hunter-Gatherers in History, Archaeology and Anthropology*. New York: Berg.

Barton, Roy F. 1919. *Ifugao Law*. University of California Publication in American Archaeology and Ethnology, no. 15.

———. 1949. *The Kalingas*. Chicago: University of Chicago Press.

Benedict, Ruth. 1934. *Patterns of Culture*. Boston: Houghton Mifflin.

Bennett Ross, Jane. 1984. "Effects of Contact on Revenge Hostilities among the Achuara Jivaro." In *Warfare, Culture and Environment*. R. Brian Ferguson (ed.). Orlando: Academic Press.

Bigelow, Robert. 1975. "The Role of Competition and Cooperation in Human Evolution." In *War: Its Causes and Correlates*. Martin A. Nettleship, R. Dale Givens, and Anderson Nettleship (eds.). The Hague: Mouton.

Bishop, Charles. 2007. "Cree-Inuit Warfare in the Hudson Bay Region." In *North American Indigenous Warfare and Ritual Violence*. Richard Chacon and Ruben Mendoza (eds.). Tucson: University of Arizona Press.

Blanchard, D. Caroline, Mark Hebert, and Robert J. Blanchard. 1999. "Continuity vs. Political Correctness: Animal Models and Human Aggression."

The HFG Review: A Publication of the Harry Frank Guggenheim Founda-tion 3(1):3–12.

Boas, Franz. 1888. *The Central Eskimo*. Bureau of American Ethnology, Annual Report 6:399–699. Washington, DC: Smithsonian Institution.

Bodley, John H. 1982. *Victims of Progress*, 2nd ed. Menlo Park, CA: Benjamin/ Cummings.

Boehm, Christopher. 1983. *Montenegrin Social Organization and Values: Political Ethnography of a Refuge Area Tribal Adaptation*. New York: AMS Press.

———. 1984. *Blood Revenge: The Anthropology of Feuding in Montenegro and Other Tribal Societies*. Lawrence: University Press of Kansas.

———. 1999. *Hierarchy in the Forest: The Evolution of Egalitarian Behavior*. Cambridge: Harvard University Press.

Bower, Bruce. 1995. "Ultrasocial Darwinism: Cultural Groups May Call the Evolutionary Shots in Modern Societies." *Science News* 48:366–367.

———. 2001. "Rumble in the Jungle: A Bitter Scientific Dispute Erupts Around the Yanomamö Indians." *Science News* 159:58–60.

Brown, Seymour. 1987. *The Causes and Prevention of War*. New York: St. Martin's Press.

Brues, Alice. 1959. "The Spearman and the Archer: An Essay on Selection in Body Build." *American Anthropologist* 61:457–469.

Carneiro, Robert. 1970. "Foreword." In *The Evolution of War: A Cross-Cultural Study*, Keith F. Otterbein. New Haven: Human Relations Area Files Press.

———. 1981. "The Chiefdom: Precursor of the State." In *The Transition to Statehood in the New World*. Grant D. Jones and Robert R. Kautz (eds.). Cambridge, MA: Cambridge University Press.

Chagnon, Napoleon. 1968. *Yanomamö: The Fierce People*. New York: Holt, Rinehart and Winston.

———. 1988. "Life Histories, Blood Revenge, and Warfare in a Tribal Popula-tion." *Science* 239:985–992.

———. 1997. *Yanomamö*, 5th ed. New York: Holt Rinehart and Winston.

Chaliand, Gèrard (ed.). 1994. *The Art of War in World History: From Antiquity to the Nuclear Age*. Berkeley: University of California Press.

Claessen, Henri J. M., and Peter Skalnik (eds.). 1978. *The Early State*. The Hague: Mouton.

Codere, Helen. 1950. *Fighting with Property: A Study of Kwakiutl Potlatching and Warfare: 1792–1930*. American Ethnological Society Monograph no. 18.

Cohen, Ronald. 1981. "Evolution, Fission, and the Early State." In *The Study of the State*. Henri J. M. Claessen and Peter Skalnik (eds.). The Hague: Mouton.

Coon, Carleton S. 1971. *The Hunting Peoples*. Boston, Little, Brown.

Corry, Stephen. 1996. "Genocide." In *The Encyclopedia of Cultural Anthropol-ogy*. David Levinson and Melvin Ember (eds.). New York: Henry Holt.

Coser, Lewis A. 1956. *The Functions of Social Conflict*. Glencoe: Free Press.

Coski, John M. 2005.*The Confederate Battle Flag: America's Most Embattled Emblem*. Cambridge, MA: Belknap Press of Harvard University Press.

Cotterall, Arthur. 2005. *Chariot: The Astounding Rise and Fall of the World's First War Machine*. Woodstock and New York: The Overlook Press.

Daly, Martin, and Margo Wilson. 1988. *Homicide*. Hawthorne, NY: Aldine de Gruyter.

Darwin, Charles. 1859. *The Origin of Species by Means of Natural Selection, or the Preservation of Favoured Races in the Struggle for Life*. London: John Murray.

Delbrück, Hans. 1975–1985. *History of the Art of War Within the Framework of Political History*, 4 volumes. Westport, CT: Greenwood Press. (German edition, 4 vols., 1904–1920.)

Dentan, Robert K. 1979[1968]. *The Semai: A Nonviolent People of Malaya*. New York: Holt, Rinehart and Winston.

———. 1992. "The Rise, Maintenance, and Destruction of Peaceable Polities: A Preliminary Essay in Political Ecology." In *Aggression and Peacefulness in Humans and Other Primates*, James Silverberg and J. Patrick Gray (eds.). New York: Oxford University Press.

Department of the Army. 1962. *Counterguerrilla Operations*. Field Manual, no. 31-16.

———. 1976. *Field Service Regulations Operations*. Field Manual, no. 100-5.

———. 2007. *Counterinsurgency*. Field Manual, no. 3-24. Chicago: University of Chicago Press.

DeQuervain, D. J.-F., et al. 2004. "The Neural Basis of Altruistic Punishment." *Science* 305:1254.

Dickson, D. Bruce. 1985. "The Atlatl Assessed: A Review of Recent Anthropological Approaches to Prehistoric North America Weaponry." *Bulletin of the Texas Archaeological Society* 56:1–38.

Dillon, Richard G. 1980a. "Capital Punishment in Egalitarian Society: The Meta Case." *Journal of Anthropological Research* 36:437–452.

———. 1980b. "Violent Conflict in Meta Society." *American Ethnologist* 7:658–673.

Divale, William T., and Marvin Harris. 1976. "Population, Warfare, and the Male Supremacist Complex." *American Anthropologist* 78:521–538.

Dollard, John, et al. 1939. *Frustration and Aggression*. New Haven: Yale University Press.

Drews, Robert. 1993. *The End of the Bronze Age: Changes in Warfare and the Catastrophe ca. 1200 BC*. Princeton: Princeton University Press.

———. 2004. *Early Riders: The Beginnings of Mounted Warfare in Asia and Europe*. Princeton: Princeton University Press.

Ducrey, Pierre. 1985. *Warfare in Ancient Greece*. Janet Lloyd, trans. New York: Schocken Books.

Ebert, C. H. V. 2000. *Disasters: An Analysis of Natural and Human-Induced Hazards*. Dubuque, IA: Kendall Hunt.

Edgerton, Robert B. 1988. *Like Lions They Fought: The Zulu War and the Last Black Empire in South Africa*. New York: The Free Press.

Eller, Jack David. 2006. *Violence and Culture: A Cross-Cultural and Interdisciplinary Approach*. Belmont, CA: Thompson/Wadsworth.

Ember, Carol. 1978. "The Myths About Hunter-Gatherers." *Ethnology* 17:439–448.

Ember, Melvin. 1982. "Statistical Evidence for an Ecological Explanation of Warfare." *American Anthropologist* 84:571–594.

————, and Carol Ember. 1971. "The Conditions Favoring Matrilocal and Patrilocal Residence." *American Anthropologist* 73:571–594.

Emerson, Robert L. 2004. "The Battle of Newtown: August 29, 1779." *Fortress Niagara* 5 (4, June):5–10.

Endicott, Kirk M., and Karen L. Endicott. 2008. *The Headman Was a Woman: The Gender Egalitarian Batek of Malaysia*. Long Grove, IL: Waveland Press.

Engles, Donald W. 1978. *Alexander the Great and the Logistics of the Macedonian Army*. Berkeley: University of California Press.

Fagan, Brian. 1999. *Floods, Famines and Emperors: El Niño and the Fate of Civilizations*. New York: Basic Books.

Feinman, Gary M. and Joyce Marcus (eds.). 1998. *Archaic States*. Santa Fe, NM: School of American Research.

Ferguson, R. Brian. 1992. "A Savage Encounter: Western Contact and the Yanomamö Warfare Complex." In *War in the Tribal Zone: Expanding States and Indigenous Warfare*. R. Brian Ferguson and Neil Whitehead (eds.). Seattle: University of Washington Press.

————. 1995. *Yanomami Warfare: A Political History*. Santa Fe, NM: School of American Research.

————. 2003. "The Birth of War." *Natural History*. July/August:28–35.

————. 2006. "Archaeology, Cultural Anthropology and the Origins and Intensifications of War." In *The Archaeology of Warfare: Prehistories of Raiding and Conquest*. Elizabeth N. Arkush and Mark W. Allen (eds.). Gainesville: University of Florida Press. Pp. 469–523.

————, and Neil Whitehead (eds.). 1992. *War in the Tribal Zone: Expanding States and Indigenous Warfare*. Seattle: University of Washington Press.

————, and Neil Whitehead (eds.). 1999. *War in the Tribal Zone: Expanding States and Indigenous Warfare*. 2nd ed. Santa Fe, NM: School of American Research.

Ferrill, Arther. 1985. *The Origins of War From the Stone Age to Alexander the Great*. New York: Thames and Hudson.

Finlayson, Clive. 2004. *Neanderthals and Modern Humans: An Ecological and Evolutionary Perspective*. Cambridge: Cambridge University Press.

Fleising, Usher, and Sheldon Goldenberg. 1987. "Ecology, Social Structure, and Blood Feud." *Behavior Science Research* 21:160–181.

Flexner, James Thomas. 1984. *Washington: The Indispensable Man*. Boston: Little, Brown.

Forde, C. Daryll. 1963. *Habitat, Economy, and Society*. New York: E. P. Dutton & Co. Inc.

Fortes, Meyer, and E. E. Evans-Pritchard. 1940. "Introduction." In *African Political Systems*, Meyer Fortes and E. E. Evans-Pritchard, eds. London: Oxford University Press.

Fox, Richard A., Jr. 1993. *Archaeology, History, and Custer's Last Battle: The Little Big Horn Reexamined*. Norman: University of Oklahoma Press.

Freud, Sigmund. 1959[1932]. "Why War?" In *Collected Papers*, J. Strachey (ed.). New York: Basic Books. Pp. 273–287.

Fried, Morton H. 1967. *The Evolution of Political Society: An Essay in Political Anthropology*. New York: Random House.

Friedman, Meyer, and Ray H. Rosenman. 1974. *Type A Behavior and Your Heart*. New York: Fawcett Crest.

Frison, George C. 1978. *Prehistoric Hunters of the High Plains*. New York: Academic Press.

———. 2004. *Survival by Hunting: Prehistoric Human Predators and Animal Prey*. Berkeley: University of California Press.

Fry, Douglas P. 2006. *The Human Potential for Peace: An Anthropological Challenge to Assumptions about War and Violence*. New York: Oxford University Press.

———. 2007. *Beyond War: The Human Potential for Peace*. New York: Oxford University Press.

Gabriel, Richard A. 2003. *The Military History of Ancient Israel*. Westport, CT: Praeger.

Ganzhorn, John W. 1959. *I've Killed Men: An Epic of Early Arizona*. New York: Devin-Adair.

Gat, Azar. 2006. *War in Human Civilization*, Oxford: Oxford University Press.

Gearing, Fred. 1962. *Priests and Warriors: Social Structures for Cherokee Politics in the 18th Century*. Memoir 93, American Anthropological Association, 64(5): part 2.

Gluckman, Max. 1974. "The Individual in a Social Framework: The Rise of King Shaka of Zululand." *Journal of African Studies* 9(2):113–144.

González-José, Rolando, Maria Cátira, Bortolini Fabrício R. Santos, and Sandro L. Bonatto. 2008. "The Peopling of America: Craniofacial Shape Variation on a Continental Scale and its Interpretation From an Interdisciplinary View." *American Journal of Physical Anthropology* 137:175–187.

Green, Thomas A. (ed.). 2001. *Martial Arts of the World: An Encyclopedia*. Santa Barbara, CA: ABC-CLIO Publishing.

Guevara, Che. 1969. *Guerilla Warfare*. J. P. Morray, trans. New York: Vintage Books/Random House.

Gulaine, Jean and Jean Zammit. 2005[2001]. *The Origins of War: Violence in Prehistory*. Melanie Hersey, trans. Malden, MA: Blackwell.

Haas, Jonathan. 1982. *The Evolution of the Prehistoric State*. New York: Columbia University Press.

———. 2001. "Warfare and the Evolution of Culture." In *Archaeology at the Millennium*. Gary M. Feinman and T. Douglas Price (eds.). New York: Kluwer Academic/Plenum. Pp. 329–350.

Harrison, Robert. 1973. *Warfare*. Minneapolis: Burgess.

Hart, C. W. M., and Arnold R. Pilling. 1960. *The Tiwi of North Australia*. New York: Holt, Rinehart and Winston.

Hassig, Ross. 1988. *Aztec Warfare: Imperial Expanse and Political Control*. Norman: University of Oklahoma Press.

Headrick, Daniel R. 1981.*The Tools of Empire: Technology and European Imperialism in the Nineteenth Century*. New York: Oxford University Press.

Heider, Karl. 1970. *The Dugum Dani: A Papuan Culture in the Highlands of West New Guinea*. Chicago: Aldine.

———. 1979. *Grand Valley Dani: Peaceful Warriors*. New York: Holt, Rinehart and Winston.

———. 1991. *Grand Valley Dani: Peaceful Warriors*. 2nd ed. New York: Holt, Rinehart and Winston.

———. 1997. *Grand Valley Dani: Peaceful Warriors*. 3rd ed. Fort Worth: Harcourt Brace.

Henderson, Charles. 1986. *Marine Sniper: 93 Confirmed Kills*. Briarcliff Manor, NY: Stein and Day.

———. 2000. *Silent Warrior*. New York: Berkley Books.

Henry, Donald O. 1989. *From Foraging to Agriculture: The Levant at the End of the Ice Age*. Philadelphia: University of Pennsylvania Press.

Herzog, Chaim, and Moedechai Gichon. 2006. *Battles of the Bible*. New York: Barnes and Noble.

Hickerson, Harold. 1965. "The Virginia Deer and Intertribal Buffer Zones in the Upper Mississippi Valley." In *Man, Culture and Animals: The Role of Animals in Human Ecological Adjustments*. A. Leeds and A. Vayda (eds.). Washington: American Association for the Advancement of Science. Pp. 43–65.

Hoebel, E. Adamson. 1940. *The Political Organization and Law-ways of the Comanche Indians*. Menasha, WI: American Anthropological Association and Contribution of the Laboratory of Anthropology.

———. 1954. *The Law of Primitive Man: A Study in Comparative Legal Dynamics*. Cambridge: Harvard University Press.

———. 1978. *The Cheyenne: Indians of the Great Plains*. 2nd ed. New York: Holt, Rinehart and Winston.

Holling, Holling C. 1935. *The Book of Indians*. New York: Platt and Munk.

Honigmann, John (ed.). 1973. *Handbook of Social and Cultural Anthropology*. Chicago: Rand McNally.

Iklé, Fred Charles. 1991. *Every War Must End* (Revised Ed.). New York: Columbia University Press.

Isserlin, B. S. J. 1998.*The Israelites*. London: Thames and Hudson.

Keegan, John. 1987. *The Mask of Command*. New York: Viking.

Keeley, Lawrence H. 1996. *War Before Civilization: The Myth of the Peaceful Savage*. New York: Oxford University Press.

Keiser, R. Lincoln. 1991. *Friend by Day, Enemy by Night: Organized Vengeance in a Kohistani Community*. New York: Holt, Rinehart and Winston.

Kelle, Brad. 2007. *Ancient Israel at War 853–586 B.C.* London: Osprey.

Kelly, Raymond C. 1985. *The Nuer Conquest: The Structure and Development of an Expansionist System*. Ann Arbor: University of Michigan Press.

———. 2000. *Peaceful Societies and the Origin of War*. Ann Arbor: University of Michigan Press.

Kennett, Lee, and James LaVerne Anderson. 1975. *The Gun in America: The Origins of a National Dilemma*. Westport, CT: Greenwood Press.

King, Barbara J. 2008. "The Anthropology of African Apes." *AnthroNotes* 29(1):1–7.

Knutson, Brian. 2004. "Sweet Revenge?" *Science* 305:1246–1247.

Kurtz, Donald V. 1978. "The Legitimation of the Aztec State." In *The Early State*. Henri J. M. Claessen and Peter Skalnik (eds.). The Hague: Mouton.

———. 2001. *Political Anthropology: Paradigms and Power*. Boulder, CO: Westview Press.

Kuschel, Rolf. 1988. *Vengeance Is Their Reply: Blood Feuds and Homicides on Bellona Island. Part 1: Conditions Underlying Generations of Bloodshed*. Copenhagen: Danks psykologisk Forlag.

Layton, Robert, and Robert Barton. 2001. "Warfare and Human Social Evolution." In *Ethnoarchaeology of Hunter-Gatherers: Pictures at an Exhibi-*

tion, K. J. Fewster and M. Zvelebil (eds.). BAR International Series 955, pp. 13–24. Oxford: Archaeopress.

Leach, Edmund. 1977. *Custom, Law, and Terrorist Violence*. Edinburgh: Edinburgh University Press.

LeBlanc, Stephan A. 2003. *Constant Battles: The Myth of the Peaceful, Noble Savage*. New York: St. Martin's Press.

Lee, Richard B. and Irven DeVore (eds.). 1968. *Man the Hunter*. Chicago: Aldine.

Lendon, J. E. 2005. *Soldiers and Ghosts: A History of Battle in Classical Antiquity*. New Haven: Yale University Press.

Long, Grahame. 2008. "Death Before Dishonor: The Fashionable yet Sometimes Fatal Ritual of Dueling in South Carolina." *Charleston Magazine* (March):112–117.

Lynn, John. 2003. *Battle: A History of Combat and Culture*. Boulder, CO: Westview Press.

Marcus, Joyce, and Gary Feinman. 1998. "Introduction." In *Archaic States*. Gary M. Feinman and Joyce Marcus (eds.). Santa Fe, NM: School of American Research.

Marcus, Joyce, and Kent V. Flannery. 1996. *Zapotec Civilization: How Urban Society Evolved in Mexico's Oaxaca Valley*. London: Thames and Hudson.

Marzke, M. W. 1983. "Joint Functions and Grips of the *Australopithecus afarensis* Hand with Special Reference to the Region of the Capitate." *Journal of Human Evolution* 12:197–211.

McFate, Montgomery. 2005. "Anthropology and Counterinsurgency: The Strange Story of Their Curious Relationship." *Military Review* (March/April).

McGrew, William C. 1992. *Chimpanzee Material Culture: Implications for Human Evolution*. Cambridge: Cambridge University Press.

Mearsheimer, John. 1983. *Conventional Deterrence*. Ithaca: Cornell University Press.

Michener, James, and A. Grove Day. 1958. "Will Mariner, the Boy Chief of Tonga." In *Rascals in Paradise*. New York: Bantam.

Michno, Gregory F. 1997. *Lakota Noon: The Indian Narrative of Custer's Defeat*. Missoula, MT: Mountain Press.

Montell, William L. 1986. *Killings: Folk Justice in the Upper South*. Lexington: University Press of Kentucky.

Moore, A. T. M., G. C. Hillman, and A. J. Legge (eds.). 2000. *Village on the Euphrates: From Foraging to Farming at Abu Hureyra*. Oxford: Oxford University Press.

Morgan, Lewis Henry. 1962[1851]. *League of the Iroquois*. New York: Corinth Press.

Morris, Donald R. 1965. *The Washing of the Spears: A History of the Rise of the Zulu Nation Under Shaka and Its Fall in the Zulu War of 1879*. New York: Simon and Schuster.

Murdock, George Peter. 1934. *Our Primitive Contemporaries*. New York: Macmillan.

———. 1959. *Africa: Its Peoples and Their Culture History*. New York: McGraw-Hill.

———. 1968a. Comment on "Are the Hunters and Gatherers a Cultural Type?" In *Man the Hunter*. Richard B. Lee and Irven DeVore (eds.). Chicago: Aldine.

———. 1968b. "The Current Status of the World's Hunters and Gatherers." In *Man the Hunter*, Richard B. Lee and Irven DeVore (eds.). Chicago: Aldine.

———. 1981. *Atlas of World Cultures.* Pittsburgh: University of Pittsburgh Press.

Neher, Gerald, and Lois Neher. n.d. *The Chibuk.* Unpublished manuscript.

Naroll, Raoul. 1964. "On Ethnic Unit Classification." *Current Anthropology* 5:283–312.

———. 1966. "Does Military Deterrence Deter?" *Trans-Action* 3(2):4–20.

———, Vern Bullough, and Frada Naroll. 1974. *Military Deterrence in History: A Pilot Cross-Historical Survey.* Albany: State University of New York Press.

Oman, Charles W. C. 1960[1885]. *The Art of War in the Middle Ages, A.D. 378–1515.* Ithaca: Great Seal Books.

Otterbein, Charlotte Swanson. 1970. "Methodological Aspects of the Military Sophistication Scale." In *The Evolution of War,* Keith F. Otterbein. New Haven: Human Relations Area Files Press. Appendix B.

———, and Keith F. Otterbein. 1973. "Believers and Beaters: A Case Study of Childrearing in the Bahamas." *American Anthropologist* 75:1670–1681.

Otterbein, Keith F. 1964a. "The Evolution of Zulu Warfare." *Kansas Journal of Sociology* 1:27–35. (Reprinted in: *Law and Warfare: Studies in the Anthropology of Conflict.* Paul Bohannan [ed.]. New York, The Natural History Press, 1967. Pp. 351–357. Also reprinted in: *Feuding and Warfare: Selected Works of Keith F Otterbein.* 1994.)

———. 1964b. "Why the Iroquois Won: An Analysis of Iroquois Military Tactics." *Ethnohistory* 11:56–63. (Reprinted in: *Law and Warfare: Studies in the Anthropology of Conflict.* Paul Bohannan, [ed.].New York: The Natural History Press, 1967, pp. 345–349. Also reprinted in: *Feuding and Warfare: Selected Works of Keith F Otterbein.* 1994.)

———. 1966. "Review of *Yoruba Warfare in the Nineteenth Century* by Ajayi and Smith. *American Anthropologist* 68:1531.

———. 1968a. "Higi Armed Combat" *Southwestern Journal of Anthropology* 24:195–213. (Reprinted in: *Feuding and Warfare: Selected Works of Keith F. Otterbein.* 1994.)

———. 1968b. "Internal War: A Cross Cultural Study." *American Anthropologist* 70:277–289. (Reprinted in: Falk, Richard A. and Samuel S. Kim [eds.], *The War System: An Interdisciplinary Approach*, Boulder, CO: Westview Press, 1980. Pp. 204–223.)

———. 1970. *The Evolution of War: A Cross Cultural Study.* New Haven: Human Relations Area Files Press.

———. 1973. "The Anthropology of War." In *Handbook of Social and Cultural Anthropology*, John J. Honigmann (ed.). Chicago: Rand McNally. Pp. 923–958. (Reprinted in: *Feuding and Warfare: Selected Works of Keith F. Otterbein.* 1994.)

———. 1977. *Comparative Cultural Analysis: An Introduction to Anthropology*, 2nd ed. New York: Holt, Rinehart and Winston.

———. 1979a. "A Cross-Cultural Study of Rape." *Aggressive Behavior* 5:425–435. (Reprinted in: *Feuding and Warfare: Selected Works of Keith F. Otterbein.* 1994.)

———. 1979b. "Huron vs. Iroquois: A Case Study of Inter-Tribal Warfare." *Ethnohistory* 26:141–152. (Reprinted in: *Feuding and Warfare: Selected Works of Keith F. Otterbein.* 1994.)

———. 1985a. *The Evolution of War: A Cross Cultural Study*, 2nd ed. New Haven: Human Relations Area Files Press.

———. 1985b. "Feuding: Dispute Resolution or Dispute Continuation?" (extensive review of *Blood Revenge* by Christopher Boehm). *Reviews In Anthropology* 12:73–83. (Reprinted in: *Feuding and Warfare: Selected Works of Keith F. Otterbein*. 1994.)

———. 1986. *The Ultimate Coercive Sanction: A Cross-Cultural Study of Capital Punishment*. New Haven: Human Relations Area Files Press.

———. 1988. "Capital Punishment: A Selection Mechanism": Comment on Robert K. Dentan, "On Semai Homicide." *Current Anthropology* 29:633–636.

———. 1989a. "The Dilemma of Disarming." In *Cold War and Nuclear Madness: An Anthropological Analysis*. Paul R. Turner, David Pitt, and Contributors, South Hadley, MA: Bergin and Garvey. Pp. 128–137. (Reprinted in: *Feuding and Warfare: Selected Works of Keith F. Otterbein*. 1994.)

———. 1989b. *The Evolution of War: A Cross-Cultural Study*, 3rd ed. New Haven: Human Relations Area Files Press.

———. 1994a. "Convergence in the Anthropological Study of Warfare." In *Feuding and Warfare: Selected Works of Keith F. Otterbein*. Langhorne, PA: Gordon and Breach. Pp. 171–180.

———. 1994b. "Ethnic Soldiers, Messiahs, and Cockalorums." (extensive reviews of *War in the Tribal Zone* by Brian Ferguson and Neil L. Whitehead, *War of Shadows* by Michael F. Brown and Eduardo Fernandez, and *Stains on My Name, War in My Veins* by Brackette F. Williams) *Reviews In Anthropology* 23:213–225.

——— (ed.).1994c. *Feuding and Warfare: Selected Works of Keith F. Otterbein*. Langhorne, PA: Gordon and Breach.

———. 1995. "More on the Nuer Expansion." *Current Anthropology* 36:821–823.

———. 1996a"Crime." In *The Encyclopedia of Cultural Anthropology*. David Levinson and Melvin Ember (eds.). New York: Henry Holt. Vol. 1, pp. 254–257.

———. 1996b. "Feuding." In *The Encyclopedia of Cultural Anthropology*. David Levinson and Melvin Ember (eds.). New York: Henry Holt. Vol. 2, pp. 493–496.

———. 1997. "The Origins of War." *Critical Review* 11:251–277.

———. 1999a. "Clan and Tribal Conflict." In *Encyclopedia of Violence, Peace, and Conflict*. Lester R. Kurtz (ed.). San Diego, Academic Press Vol. 1, pp. 289–295.

———. 1999b. "A History of Research on Warfare in Anthropology." *American Anthropologist* 101:794–805.

———. 2000a. "The Doves Have Been Heard from, Where Are the Hawks?" *American Anthropologist* 102:841–844.

———. 2000b. "Five Feuds: An Analysis of Homicides in Eastern Kentucky in the Late Nineteenth Century." *American Anthropologist* 102:231–243.

———. 2000c. "The Killing of Captured Enemies: A Cross-Cultural Study." *Current Anthropology* 41(3):439–443.

———. 2001. "Dueling." In *Martial Arts of the World: An Encyclopedia*. Thomas A, Green (ed.). Santa Barbara, CA: ABC-CLIO. Vol. 1, pp. 97–108.

———. 2004. *How War Began*. College Station, TX: Texas A & M University Press.

———. 2005a. "Hunting and Virilocality." *Current Anthropology* 46:124–127.

———. 2005b. "Preface." In *The Balance of Human Kindness and Cruelty: Why We Are the Way We Are,* Robert B. Edgerton. Lewiston, NY: The Mellen Press. Pp. xiii–xv.

———. 2006a. "Feuds and Feuding." In *The New Encyclopedia of Southern Culture. Vol. 4: Myth, Manners, and Memory.* Charles Wilson (ed.). Chapel Hill: University of North Carolina Press. Pp. 225–227.

———. 2006b. "War, Archaeology of." In *Encyclopedia of Anthropology.* H. James Birx (ed.). Thousand Oaks, CA: Sage. Pp. 2300–2302.

———. 2008. "Clan and Tribal Conflict." In *Encyclopedia of Violence, Peace, and Conflict,* 2nd ed. Lester R. Kurtz (ed.). San Diego: Academic Press.

———, and Charlotte Swanson Otterbein. 1965. "An Eye for an Eye and a Tooth For a Tooth: A Cross-Cultural Study of Feuding." *American Anthropologist* 67:1470–1482.

Perlmutter, David D. 1999. *Visions of War: Picturing Warfare from the Stone Age to the Cyber Age.* New York: St. Martin's Press.

Persig, Robert M. 1974. *Zen and the Art of Motorcycle Maintenance: An Inquiry into Values.* New York: William Morrow.

Peters, Ralph. 1999. *Fighting for the Future: Will America Triumph?* Mechanicsburg, PA: Stackpole.

Peterson, Roger Tory. 1980. *A Field Guide to the Birds: A Completely New Guide to All the Birds of Eastern and Central North America.* Boston: Houghton Mifflin.

Pilling, Arnold R. 1968. "Discussion: Predation and Warfare." In *Man the Hunter.* R. B. Lee and I. DeVore (eds.). Chicago: Aldine.

———. 1988. "Sneak Attacks." In *The Tiwi of North Australia*, 3rd ed. C. W. M. Hart, Arnold R. Pilling, and Jane C. Goodale (eds.). New York: Holt, Rinehart and Winston.

Putnam, P. T. L. 1948. "Pygmies of the Ituri Forest." In *A Reader in General Anthropology,* Carleton S. Coon (ed.). New York: Holt. Pp. 322–342.

Rappaport, Roy A. 1968. *Pigs for the Ancestors: Ritual in the Ecology of a New Guinea People.* New Haven: Yale University Press. (Reissued: Long Grove, IL: Waveland Press, 2000.)

Peters, Ralph. 1999. *Fighting for the Future: Will America Triumph?* Mechanicsburg, PA: Stackpole Books.

Rice, Otis K. 1978. *The Hatfields and the McCoys.* Lexington: University Press of Kentucky.

Richardson, Peter, and Robert Boyd. 1998. "The Evolution of Human Ultrasociety." In *Indoctrinability, Ideology, and Warfare.* Irenaus Eibl-Eibesfeldt and Frank Kemp Salter (eds.). New York: Berghahn Books.

Richter, Werner. 1943. "The War Pattern of Swiss Life." *Foreign Affairs* 22:643.

Ritter, E. A. 1957. *Shaka Zulu: The Rise of the Zulu Empire.* New York: G. P. Putnam's Sons.

Ritter, Gerhard. 1968. *Frederick the Great: A Historical Profile.* Peter Perel, trans. Berkley: University of California Press.

Robarchek, Clayton, and Carole Robarchek. 1992. "Culture of War, Culture of Peace: A Comparative Study of Semai and Waorani." In *Aggression and Peacefulness in Humans and Other Primates.* J. Silverberg and J. A. Gray (eds.). Oxford: Oxford University Press. Pp. 189–213.

———. 1998. *Waorani: The Contexts of Violence and War*. New York: Harcourt Brace.

Rodseth, Lars, and Richard Wrangham. 2004. "Human Kinship: A Continuation of Politics by Other Means." In *Kinship and Behavior in Primates*. Bernard Chapais and Carol M. Berman (eds.). Oxford: Oxford University Press. Pp. 389–419.

Ross, Marc Howard. 1993. *The Conflict of Culture*. New Haven: Yale University Press.

Schelling, Thomas C. 1966. *Arms and Influence*. New Haven: Yale University Press.

Schneider, Joseph. 1950. "Primitive Warfare: A Methodological Note." *American Sociological Review* 15:772–777.

Scott, Douglas D. and Richard A. Fox, Jr. 1987. *Archaeological Insights into the Custer Battle: An Assessment of the 1984 Field Session*. Norman: University of Oklahoma Press.

Scott, Robert L., Jr. 1943. *God is My Co-Pilot*. New York: Scribner.

Secoy, Frank R. 1953. *Changing Military Patterns on the Great Plains*. American Ethnological Society monograph no. 21.

Service, Elman R. 1962. *Primitive Social Organization: An Evolutionary Perspective*. New York: Random House.

———.1963. *Profiles in Ethnology* (A revision of *Profiles in Primitive Culture*). New York: Harper and Row.

———. 1979. *The Hunters*. 2nd ed. New York: Prentice-Hall.

Shea, John J. 1997. "Middle Paleolithic Spear Point Technology." In *Projectile Technology*. Heidi Knecht (ed.). New York: Plenum Press.

———. 2006. "The Origins of Lithic Projectile Point Technology: Evidence from Africa, the Levant, and Europe." *Journal of Archaeological Science* 33:823–846.

Sherif, Musafer, et al. 1961. *Intergroup Conflict and Cooperation: The Robbers Cave Experiment*. Norman: University of Oklahoma, Institute of Group Relations.

Shields, Dean. 1980. "A Comparative Study of Human Sacrifice." *Behavior Science Research* 15:245–262.

Shipman, Pat. 2006. "What Do New Discoveries Tell Us About Human Evolution?" *Teaching Anthropology: SACC Notes* 12(2): 9–12.

Sipes, Richard G. 1973. "War, Sports and Aggression: An Empirical Test of Two Rival Theories," *American Anthropologist* 75:64–86.

Southwaite, David. 1984. *Battlefields of Britain*. Exeter: Wolf and Bower.

Spencer, Walter B., and F. J. Gillen. 1927. *The Arunta: A Study of a Stone Age People*. London: Macmillan.

Spitzer, Steven. 1979. "Notes Toward a Theory of Punishment and Social Change." *Research in Law and Sociology* 2:207–229.

Sponsel, Leslie. 1998. "Yanomamö: An Arena of Conflict and Aggression in the Amazon." *Aggressive Behavior* 24(2):97–122.

Sumner, William Graham. 1906. *Folkways*. Boston: Ginn.

Swanson, Guy. 1960. *The Birth of the Gods*. Ann Arbor: The University of Michigan Press.

Tacon, Paul S., and Christopher Chippendale. 1994. "Australia's Ancient Warriors: Changing Depictions of Fighting in the Rock Art of Arnhem Land, N. T." *Cambridge Archaeological Journal* 4:211–248.

Taleb, Nassim Nicholas. 2005. *Fooled by Randomness: The Hidden Role of Chance in Life and in the Markets*, 2nd ed. New York: Random House.

———. 2007. *The Black Swan: The Impact of the Highly Improbable*. New York: Random House.

Thoden van Velzen, H. U. E., and W. van Wetering. 1960. "Residence, Power Groups and Intrasocietal Aggression." *International Archives of Ethnography* 49:169–200.

Thomas, Elizabeth M. 1959. *The Harmless People*. New York: Knopf.

Turney-High, H. H. 1994[1949]. *Primitive War: Its Practice and Concepts*. Columbia: University of South Carolina Press.

Tylor, Edward B. 1888. "On a Method of Investigating the Development of Institutions; Applied to Laws of Marriage and Descent." *Journal of the Royal Anthropological Institute of Great Britain and Ireland* 18:245–270.

UN (United Nations). 1946. "The Crime of Genocide." *Resolutions Adopted by the General Assembly During Its First Session* 96(1):188–189. (http://www.un.org/documents/ga/res/1/ares1.htm)

van Beek, Walter E. A. 1987. *The Kapsiki of the Mandara Hills*. Long Grove, IL: Waveland Press.

Vayda, Andrew P. 1961. "Expansion and Warfare among Swidden Agriculturalists." *American Anthropologist* 63:346–358.

———. 1968. "Hypotheses About Functions of War." In *War: The Anthropology of Armed Conflict and Aggression*. M. Fried, M. Harris, and R. Murphy (eds.). Garden City: The Natural History Press.

———. 1976. *War in Ecological Perspective*. New York: Plenum Press.

Virga, Vincent and the Library of Congress. 2007. *Cartographia: Mapping Civilizations*. New York: Little, Brown.

Vogel, Gretchen. 2004. "The Evolution of the Golden Rule." *Science* 303:1128–1131.

Wagnespack, Nicole. 2005. "What Do Women Do in Large-Game Hunting Societies?" *American Anthropologist* 107:666–676.

Warner, W. Lloyd. 1931. "Murngin Warfare." *Oceania* 1:457–494.

———. 1964[1937]. *A Black Civilization*. New York: Harper.

Wilson, Edward O. 1975. *Sociobiology: The New Synthesis*. Cambridge: Harvard University Press.

———. 1978. *On Human Nature*. Cambridge: Harvard University Press.

Wintringham, Tom. 1943. *The Story of Weapons and Tactics from Troy to Stalingrad*. Boston: Houghton Mifflin.

Wolf, Eric. 1982. *Europe and the People without History*. Berkeley: University of California Press.

Woodburn, James. 1982. "Egalitarian Societies." *Man* (N.S.) 17:431–451.

Wrangham, Richard W. 1999. "Evolution of Coalitional Killing." *Yearbook of Physical Anthropology* 42:1–30.

Yadin, Yigail. 1963. *The Art of Warfare in Biblical Lands: In the Light of Archaeological Study*. 2 vols. New York: McGraw-Hill.

Younger, Stephen M. 2005. *Leadership, Violence, and Warfare in Small Societies: A Simulation Approach*. Los Alamos: Los Alamos National Laboratory.

———. 2008. "Conditions and Mechanisms for Peace in Precontact Polynesia. *Current Anthropology* 49:927–934.

Index